LEPIDOPTERA

Edited by **Farzana Khan Perveen**

Lepidoptera
http://dx.doi.org/10.5772/intechopen.68337
Edited by Farzana Khan Perveen

Contributors

Diego F. Villanueva-Mejía, Javier Correa Alvarez, Viviana Ramírez-Ríos, Weibin Jiang, Joji Otaki, Mayo Iwasaki, Ramón Rebolledo, Hernán Navarrete, Jurate De Prins, Farzana Khan Khan Perveen

Published by InTech
Janeza Trdine 9, 51000 Rijeka, Croatia

Publishing Process Manager Renata Sliva
Technical Editor SPi Global
Cover InTech Design team

First published November, 2017
Printed in Croatia
Legal deposit, Croatia: National and University Library in Zagreb

Additional hard copies can be obtained from orders@intechopen.com

Lepidoptera, Edited by Farzana Khan Perveen
p. cm.
Print ISBN 978-953-51-3659-0
Online ISBN 978-953-51-3660-6

For EU product safety concerns:
IN TECH d.o.o., Prolaz Marije Krucifikse Kozulić 3, 51000 Rijeka, Croatia,
info@intechopen.com or visit our website at intechopen.com.

Meet the editor

Dr. Farzana Khan Perveen (FLS; gold medalist) obtained her BSc (Hons) and MSc (Zoology—Entomology) degrees from the University of Karachi; MAS (Monbusho Scholar; Agriculture—Agronomy) from the Nagoya University, Japan; and PhD (Research and Courseworks from the Nagoya University; Toxicology) degree from the University of Karachi. She is the founder/chairperson of the Department of Zoology (DOZ) and ex-controller of examinations at Shaheed Benazir Bhutto University (SBBU) and ex-founder/ex-chairperson of DOZ, Hazara University and Kohat University of Science and Technology. She is the author of 150 high-impact research papers, 135 abstracts, 40 authored books, 5 chapters and editor of 5 books. She has supervised BS (4), MSc (50), MPhil (40), and PhD (1) students and organized and participated in numerous international and national conferences and received multiple awards and fellowships. Dr. Farzana Khan Perveen is a member of research societies, editorial boards of journals, and World Commission on Protected Areas, International Union for Conservation of Nature. Her fields of interest are entomology, toxicology, forensic entomology, and zoology.

Contents

Preface

Lepidoptera are beautiful creatures of nature. Their diversity, bright colors, patterns, marvelous shapes of wings, and graceful flight catch the attention of evolutionary biologists and give pleasure to everyone. The word Lepidoptera stands for insects with scaly wings. It is the second largest, widely spread, and recognized insect order in the world. Furthermore, approximately 20,000 species of butterflies and 180,000 species of moths, including micro- and macromoths, are found at the present. They comprise ca. 10% of the total volume of described species of living organisms. In several Lepidoptera, a period of suspended development, i.e., diapause, is prolonged or lasts more than a year. They are beneficial as well as harmful with great aesthetic and commercial values and are subject to ecological significance. They are found in all environments and provide many essential and economically imperative services within different ecosystems. They act as bioindicators, nutrient recyclers, pollinators, seed dispersal agents, and soil-formation agents. Additionally, they provide the best food chain for natural predators. For researchers, scientists, and students, they offer a model taxon to study intensively on biodiversity, conservation studies, environmental impact estimates, monitoring of animal populations, ecology, ethnology, evolution, genetics, and systematic and many other ecological and genetic studies. They open doors to establishment of chemical ecology as a scientific discipline for study and research.

The diversity of Lepidoptera is one of the most fascinating subjects of biology. Evolution, natural selection, and many other biotic and abiotic factors have produced different species of Lepidoptera, and the speciation process is continuously going on. They are common in smaller or larger institutional collections, and they are disproportionally abundant in private collections. A rough estimation of the Lepidoptera specimens in the depositories worldwide could reach about 10% from the estimated 2.5 billion of natural history collection objects. The information science and data management tools became very important in the Lepidoptera collection curation. The complexity of techniques and computing tools used in taxonomy and increase in the amount of data that can be obtained from collection-based disciplines make it necessary to automatize processes in data gathering, manipulation, analysis, and visualization. In this sense, the collaboration between researchers, taxonomists, citizens, scientists, collection curators, and computing/information science is crucial to build and to use the proper approaches in taxonomy needed to avoid error. The modern approaches toward the Lepidoptera collection and data management help to focus on the goals and studies, which can be finalized.

As the supplementary to morphological research, analysis of DNA has been widely used in the phylogenetic and taxonomic studies of Lepidoptera. A conservative estimation is that the *Wolbachia Hertig* and Burt 1924 bacteria infection rate in the grass skippers, *Polytremis nascens* (Leech, 1893) (Lepidoptera, Hesperiidae), is 31% and no significant difference in the prevalence is found between the sexes. The molecular characteristics of *Wolbachia* strains were also identified in the grass skipper, *Polytremis fukia* (Evans, 1940), from seven locations in China. Three portions of the mitochondrial DNA were sequenced from the individuals of *P. fukia* to infer the effect of *Wolbachia* on host mitochondrial variation.

Eyespot color patterns in butterfly wings are formed by color-pattern determination system. This system consists of two stages of induction. The early stage is a dissipation of signals from their source, and the late stage is essentially a reaction-diffusion mechanism that involves short-range activation and long-range inhibition. Calcium (Ca) signals play an important role in color-pattern determination system in the late stage of the induction. On the basis of a linear relationship between scale size and cell size, and a relationship between scale color and scale size, the putative morphogenic signals from organizers are polyploidy signals that determine cellular size via polyploidization. Calcium signals play an important role in polyploidization. The signals that determine the final scale color of a given scale cell are highly dynamic. The large double-focus fusion eyespot on the hind wing of the peacock pansy, *Junonia almana* (Linnaeus, 1758), has been taken as an example to test the involvement of the proposed signal interactions. Using the forewing-lift method immediately after pupation, a small stainless steel ball was placed on the prospective major eyespot or background of the developing dorsal hind wing to cause a wing epithelial distortion, resulting in deformation of the major eyespot. When the exposed dorsal hind wing was covered with a piece of plastic film or placed on a surface of glass slide, adhesive tape, or silicone-coated glassine paper, the major eyespot was effectively reduced in size without a direct contact with the covering materials.

The complete mitochondrial genome of the American potato tuberworm, *Tecia solanivora* (Povolný, 1973) (Lepidoptera, Gelechiidae), is presented as a model to understand how to characterize and study a mitogenome in insects. It was sequenced, analyzed, and compared with other lepidopteran insects. The phylogenetic relationships of nine clades of the order Lepidoptera were developed using Bayesian and maximum likelihood inference, which provided well-supported results compared with other phylogenies based on both molecular and morphological traits. The utility of all information presented is to improve scientific databases and support the determination of Lepidoptera population genetic studies in the future.

Considering that butterflies (Clade, Rhopalocera) are sensitive to physical and climatic changes, e.g., of temperature, humidity, and solar radiation, produced by disturbances in their habitat, a survey of this group was carried out in a small remnant of native forest (Rucamanque) in the central valley of the Araucanía Region of Chile. The object was to record the composition, abundance, and diversity of Rhopalocera in grassland, forest, and the ecotone between them during spring and summer. The study recorded 1190 individual butterflies belonging to 4 families, 8 subfamilies, 18 genera, and 25 species. The highest values of species richness and abundance were obtained in the summer, 25 species and 953 individuals; in the spring, 9 species were recorded with a total of 237 individuals. The greatest diversity and homogeneity were found in the ecotone habitat, these environments being less diverse and more homogeneous. The greatest taxocenotic similarity was found between grassland and the ecotone; the least similarity appeared between the ecotone and forest. The greatest biocenotic similarity was found between the ecotone and forest, and the lowest correspondence between grassland and forest. Recording the species diversity of the zoological group studied in forest remnants is of great importance, both for their ecological value and in order to take informed decisions in management and planning for conservation, study, and tourism.

Dr. Farzana Khan Perveen FLS
Founder Chairperson and Associate Professor
Ex-controller Examination
Shaheed Benazir Bhutto University (SBBU)
Khyber Pakhtunkhwa, Pakistan

Section 1

Lepidoptera Systematics

Chapter 1

Introductory Chapter: Lepidoptera

Farzana Khan Perveen and Anzela Khan

Additional information is available at the end of the chapter

http://dx.doi.org/10.5772/intechopen.70452

1. Lepidoptera

The word Lepidoptera comes from the Latin word, equivalent to lepido- and from the ancient Greek words lepis and pteron mean scales and wings, respectively. Therefore, it stands for insects with scaly wings. It is the second largest, diverse, widespread, and widely recognized insect's order in the class Insecta of phylum Arthropoda. Linnaeus (1707–1778) divides it into three groups: (1) butterflies, (2) skippers, and (3) micro- and macro-moths. It consists of 126 families and 46 superfamilies (**Table 1**). They can be differentiated on morphological, anatomical, behavioral, and ecological characteristics [1]. Further, 500,250 species of Lepidoptera are described, with 70,820 species of butterflies [2, 3] and 3700 species of skippers globally [4–6]. Furthermore, about 165,000 species of moths, including micro- and macro-moths, are found up to now [6–9]. In nature, Lepidoptera regard as the symbol of beauty and grace. They are very beautiful creatures of nature (**Figure 1A**) [10–12].

1.1. Morphology

The group Rhopalocera is related to butterflies and skippers; however, Heterocera is to micro- and macro-moths. The Lepidoptera show a great diversity in forms, size, structure, and other distinctiveness (**Figure 1A** and **B**) [13, 14].

1.1.1. Head segment

Lepidoptera's head capsule is the feeding and sensory center. It is small, round, or elliptical and sclerotizes organization. The upper-middle portion of the head is called the frons; below is the clypeus, and below it is the labrum, to both sides of which the edges of the mandibles with different aspects of the maxillary palps may expand beyond and/or underneath, even when view them from front [15]. As a whole, the shape and size of the head capsule, color patterns, and location of hairs on the head are supportive in identifying species of caterpillars with aid of a microscope (**Figure 1C**).

Kingdom: Animalia

Subkingdom: Invertebrata

Super-division: Eumetazoa

Division: Bilateria

Subdivision: Ecdysozoa

Superphylum: Tactopoda

Phylum: Arthropoda von Siebold, 1848

Subphylum: Atelocerata

Superclass: Hexapoda

Class: Insecta

Infraclass: Neoptera

Subclass: Pterygota

Superorder: Endopterygota

Unranked: Amphiesmenoptera

Unranked: Holometabola

Order: Lepidoptera Linnaeus, 1758

Examples:

- Butterflies
- Skippers
- Micro-moths
- Macro-moths

Table 1. The taxonomic position of the order Lepidoptera [13].

1.1.1.1. Mouthparts

The siphoning-type mouthparts are found in the imago. They are transformed into a long flexible hollow structure, in which the formation of the suctorial proboscis encompasses a fluid-tight food tube. Lepidoptera feed on nectar, and their proboscis length may increase almost 100-folds. Usually, when they do not use them, they keep coil under the head with the help of small muscles present there. In all Lepidoptera, the basic structure of mouthparts is the same, which include each one labium, labrum, hypopharynx, or tongue with pairs of mandibles and maxillae (**Figure 1C**) [16, 17].

1.1.1.2. Antennae

The antennae show a wide variation in forms, size, structure, and other characteristics among species and even between different sexes. The basic structure of antennae of Lepidoptera is usually filiform, which is altered into the capitates of antenna, which are club shaped with a long shaft and a bulb at the end. In the skippers, most of the antennae's tips are changed

http://dx.doi.org/10.5772/intechopen.70452

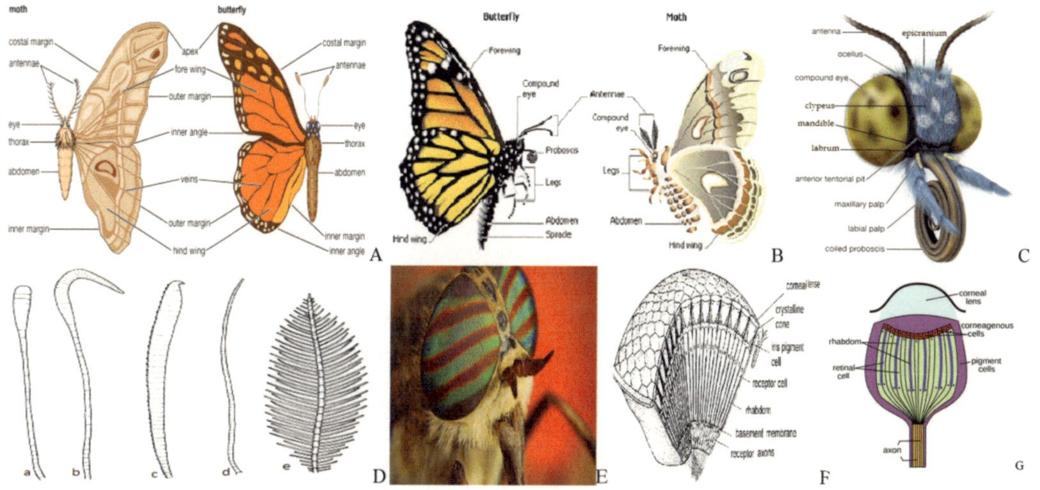

Figure 1. Parts of Lepidoptera body: (A and B) butterfly and moth; (C) generalize mouth parts; (D) generalize antennae of order Lepidoptera: butterflies; (a) skippers; (b) micro- and macro- moths (c, d, e) [18, 19]; (E) compound eye of Lepidoptera; (F) unit and basic structure, ommatidium of a compound eye, showing its construction; (G) simple eye or ocellus [22].

into a narrow hook-like projection. A moth's antennae are feathery or saw edged. They also act as the balancing organ. The shapes of antennae are assisting in identifying species of Lepidoptera (**Figure 1D**) [20].

1.1.1.3. Compound eyes

The eyes are usually paired, golden brown, or even red as in some skipper species. They built quite differently from the vertebrate eye, but like Arthropod, it is made up of repeating units up to 17,000, the ommatidia, each function as a separate visual receptor, which in combination provide a broad mosaic view of the image, such type of the eye is a compound eye. Each is connected to a lens, which is attached to a nerve leading to the brain (**Figure 1E** and **F**) [21].

1.1.1.4. Simple eyes or ocelli

In all Lepidoptera, in addition compound eyes, simple eyes, or ocelli (singular: ocellus; simple photoreceptors) are also present. They made with a single lens and several sensory cells. Only two ocelli are present in imago, excluding a few moths, one on each side of compound eyes. In some species, a type of sense organs, which called chaetosemata, is found near the ocelli. In caterpillars, three pairs of simple eyes are found, which are not homologous to ocelli of imago. Simple eyes of caterpillars are differently named as stemmata. Lepidoptera are able to perceive ultraviolet (UV) light and observe wing colors and patterns by ocelli (**Figure 1G**) [22].

1.1.2. Thorax

The thorax is the second part of the body, which composed of three jointed segments, the prothorax, mesothorax, and metathorax, each derives from a primitive segment. They are covered dorsally with tergites, ventrally with sternites, and laterally with pleurites (chitinous plates) [23]. The characteristic like the presence or absence and shape of sclerotized plates; location of primary setae; and location, color, and shape of the prothoracic spiracle assists in identification of caterpillar and imago species [24].

1.1.2.1. *Jointed legs*

Lepidoptera have three pairs of well-developed jointed legs. They are located in each segment of the thorax and covered with scales. Each leg consists of nine segments, that is, coxa, trochanter, femur, tibia; five tarsal segments with a pretarsus; and a pair of articulated curved claws on the fifth segment. Morphology of the legs also aids in identifying caterpillar and imago species (**Figure 2a** and **b**) [25]. The aroliar pad (a pad extending between the tarsal claws) and pulvillus (plural: pulvilli, pads beneath each tarsal claw) are short or absent in some families. The tibia of each leg contains a subgenual organ, which detects and amplifies small vibrations (**Figure 2a** and **b**) [26].

1.1.2.2. *Wings*

The mysterious flight of Lepidoptera depends on their wings, which accomplished several kinds of difficult tasks like diving, circling, parachuting, equilibrium, etc. all because of their lightness in nature. Their wings are subjected to considerable variations in shape, size, markings, spots, and vein patterns, thus reflecting their specific functional differences. Strong muscles in the thorax move the wings up and down in a digit 8 pattern during the flight. Both pairs of wings are covered with thousands of bright, colorful, and dull scales. Due to the presence of scales on the wings, the term for order Lepidoptera has been coined (**Figure 2f** and **g**). Wing scales adapt

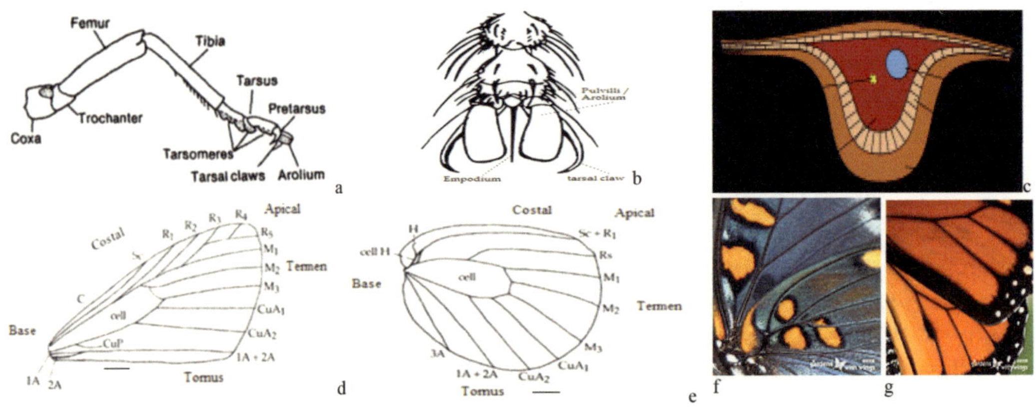

Figure 2. (a) Generalize structure of Lepidoptera Leg; (b) generalize structure of tarsus; (c) TS of vein from wing generalized wings venation: (d) forewing; (e) hindwing; with names of different veins with their abbreviations are: C: Costal; Sc: subcostal vein; R1: 1st radius vein; R2: 2nd radius vein; R3: 3rd radius vein; R4: 4th radius vein; R5: 5th radius vein; M1: 1st median vein; M2: 2nd median vein; M3: 3rd median vein; Cu1: 1st cubitus vein; CuA1: 1st cubitus anterior; CuP2: 2nd cubitus posterior; 1A: 1st anal vein; 2A: 2nd anal vein; 3A: 3rd anal vein; H: humeral 2012; bars in photographs indicate 30 mm; (f) wing color of black and white; (g) wing of a male the monarch butterfly, *Danaus plexippus* (Linnaeus, 1758) (Nymphalidae: Nymphalinae) [23–26].

to different environmental conditions affecting flights. They are also found to influence by the time factor, which affects speed, foraging, calling, finding places for spawning, avoiding predators, etc. Wing pattern is successful in establishing the correlation in adaptation and adaptive change. Both changes in morphology of wings and the pattern of flight are closely associated with the genetic material. Similarly, the relationship has been established between wing venation and wing pattern in the genus *Micropterix* Hubner, 1825 (Family: Micropterigidae). Such type of studies plays a major role in understanding its evolutionary status and highlights its significance in taxonomical analysis. They are chitinous membranes, nourished and supported by tubular veins (**Figure 2c–e**). These veins also function in exchange of oxygen. Further, venation has aerodynamic importance, plays specific role in flight system, and adapts to different surroundings [26]. Bashar et al. [27] prepared pictorial key for identification of the local nymphalid butterflies of Bangladesh, based on the wing venation of the butterflies. It has also observed that venation is an important trait in Lepidoptera phylogenetic development [28].

1.1.3. Abdomen

The abdomen is third part of the body. It consists of 10–11 segments tapering to the end. Each segment provides membranes in between allowing for articulation and movement. In Lepidoptera systematic, they have been one of the most important sources of character information [20]. In some caterpillar, four pairs of prolegs normally located on the third to sixth segments, and a separate pair of prolegs by the anus, which has a pair of tiny hooks called crotchets, helps in gripping and walking [29].

1.1.4. Scales

The name of this order Lepidoptera is due to the presence of the scales. The head, thorax, abdomen, wings, and legs are covered with minute scales, are lamellar or blade-like, and are attached with a pedicel, while other forms may be hair-like or specialized as the secondary sexual characteristics. Either, they give color, by color pigments they contain or through structural coloration with mechanism that include photonic crystal and diffraction grating. They are functioned as aiding gliding flight, insulation, producing pheromone, thermoregulation, etc. The most important is the large diversity of their bright or indistinguishable pattern, which aids the organisms to protect itself by camouflage or mimicry including rival and potential mate (**Figure 2f** and **g**) [4].

1.1.5. Sound-producing organs

The sounds of some Lepidoptera are clearly audible, for example, members of Sphingidae and Pyralidae. Many moths have developed ears on their wings or thorax, which are called as tympanal organ, which can make them aware of thread, predator, etc. The Neotropical tiger moth, *Bertholdia trigona* (Grote, 1879) (Noctuoidea: Erebidae), actively makes out of function of the bat radar by creating its own ultrasound and by vibrating its tympanum present on its metathorax. Males of the moth, *Symmoracma minoralis* (Snellen) (Pyralidae: Nymphulinae), produce a high-intensity calling song from tymbals like structure found in the genital segment [30–32].

1.2. Endocrine control

In Lepidoptera, the growth and metamorphosis are under control by interacting sets of hormones. Ecdysis is initiated by ecdysiotropin, or prothoracicotropic hormone (PTTH) or brain hormone (BH) is secreted by protocerebrum which acts on ecdysial glands. Eclosion hormone is secreted by brain median neurosecretory cells; it is stored in the corpora cardiac and is released into the hemolymph during switchover from pupa to imago. Within the abdominal ganglia, it acts on neurons to begin the pre-eclosion behavior. The corpora allata secret juvenile hormone (JH). It is liable to bring juvenile progress and variable species to species. The molting hormone (MT) or ecdysone hormone (EH) is secreted by the ecdysial gland. Distinctively, all hormones and neurohormones are involved in management of circadian rhythms, growth, development of the nervous system, diapause, mating, metabolism, oviposition, pheromone biosynthesis, regulation of dormancy, regulation of migratory behavior, and other physiological functions in the body. The EH is responsible for several activities of pupal-imago conversion, with respect to the behavior associated with ecdysis, following deterioration of abdominal intersegmental muscles. Ecdysis-triggering hormone which is the most newly discovered hormone shows a significant role in ecdysis. Bursicon (tanning hormone) is usually synthesized in neurohemal organs related with the ventral chain ganglia. Its functions are to stimulate sclerotization and tanning of the cuticle during the course of ecdysis (**Figure 3**) [33–36].

1.3. Polymorphism

Existence of the morphologically different individuals in the life cycle of the same species is termed as polymorphism. The sexual dimorphism is very common, in which male and female are structurally different and are found in families Pieridae, Nymphalidae, Papilionidae, and Psychidae. The other types are geographical, seasonal, genetic, and environmental polymorphism. In some species, the polymorphism is limited to one sex, typically the female [33–36].

1.4. Pheromones

They are biochemicals, meant of communication, secreted into the surroundings, and affect the activities or functioning in others species. Such communication systems have provided challenges to scientists in different disciplines including behavior, biochemistry, chemistry, ecology, genetics, and physiology. They produce in greater varieties. As a result, the great numbers of researches are conducted focusing on butterflies and moths. Pheromone systems are species specific for attracting mates. Sex pheromones are used for long-distance biochemical communication [37, 38].

1.5. Migration

Lepidoptera contribute an essential part as the environmental indicators. If any minute dangerous variations occur in the surroundings, they may affect acutely. Due to unfavorable environmental affects, they migrate rapidly in long distances, from locations to the area, which are more suitable for any part of the seasons. Their destination may be tropical and subtropical areas and all continents, excluding Antarctica. They stay away from undesirable

http://dx.doi.org/10.5772/intechopen.70452

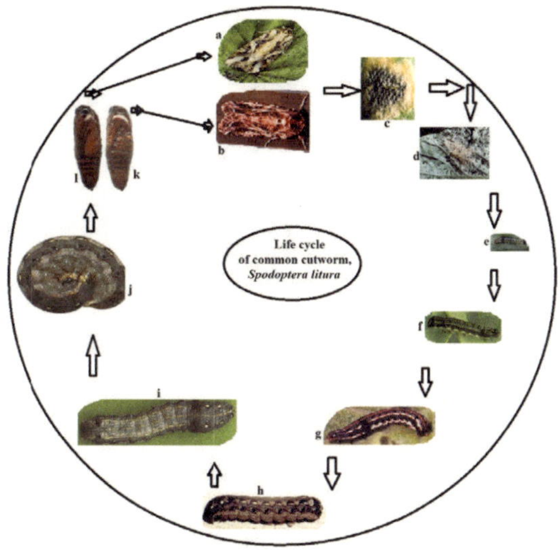

Figure 3. Life-cycle of Lepidoptera, the common cutworm, *Spodoptera litura* (Fabricius, 1775): (a) imago male; (b) imago female; (c) bunch of egg; (d) newly hatched 1st instar caterpillar; (e) 2nd instar caterpillar; (f) 3rd instar caterpillar; (g) 4th instar caterpillar; (h) 5th instar caterpillar; (i) 6th instar caterpillar; (j) defensive posture of 6th instar caterpillar; (k) female cocoon; (l) male cocoon [19].

circumstances, including adverse climate, food shortage, overpopulation, weather, etc. In some conditions, few members migrate, and in other conditions, all migrate [39–41].

1.6. Internal anatomy

1.6.1. Digestive system

In Lepidoptera, the foregut is started from the mouth; the pharynx may be highly modified into a pump. Posterior to it is the esophagus which opens into a crop or storage organ. Immediately, posterior to it is the proventriculus, a structure that contains sclerotized toothlike denticles, aiding in grinding the food. Some fluid-feeding Lepidoptera lack a proventriculus. The stomodael valve (foregut valve) regulates the flow of materials from the foregut-midgut. The former is lined with chitinous protective layer called the intima. It prohibits absorption of nutrients. In many Lepidoptera, the foregut valve has associated with gastric caeca that produce digestive enzymes and increase surface area. The intima is absent in the midgut, and most of the absorption of nutrients occurs here. The Malpighian tubules attach to the pylorus region of the hindgut. Posterior to them is the anterior intestine and the highly muscularized rectum that terminates in the anus. It functions in removing water from the fecal materials; therefore, Lepidoptera produce very dry excrements (**Figure 4**) [42–45].

1.6.2. Circulatory system

Lepidoptera have open circulatory system. The major portion of the hemolymph is found in open cavities. It bathes the organs within the body cavity, the hemocoel. Hemolymph enters

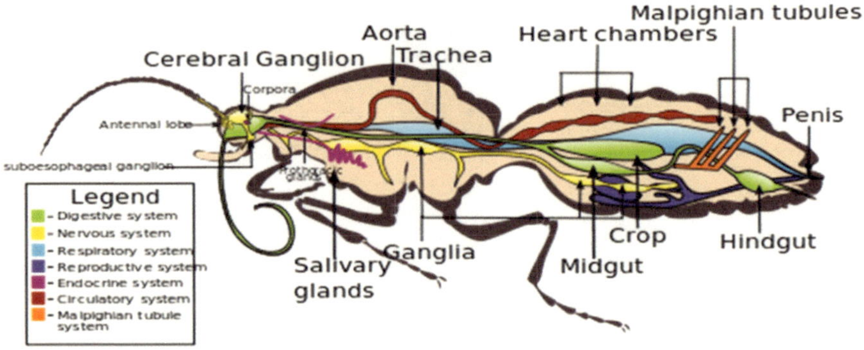

Figure 4. Internal anatomy of Lepidoptera viewing imago male (family: Nymphalidae), showing the most of the major organ systems, with characteristic reduced forelegs of that family and the corpora include the corpus allatum and the corpus cardiac [41].

the dorsal vessel or heart via small openings called the ostia. It is then pumped toward the head, where it then returns to the hemocoel. It serves as a lubricant for the movement of internal structures. It is a hydraulic medium for applying pressure for molting, eversible glands are extruded via pressure changes, and some muscular contraction is opposed by hydrostatic pressure within the hemocoel. Hemolymph transports various substances from one tissue to another (**Figure 4**) [42–45].

1.6.3. Respiratory or tracheal system

Tracheal system consists of a system of branching tubes (tracheae) and openings to the outside called spiracles. Atria spiracles have mechanisms that allow Lepidoptera to close the opening. It prevents water loss and prevents the entry of pathogens and parasites. Tubes begin rather large and branch to become successively smaller and smaller as they penetrate deep within the tissues of the insect. The smallest branches of the tracheae are the tracheoles, and gaseous exchange occurs here. Air sacs have many potential functions, all rather speculative, including increasing the volume of air in the body for exchange, lowering the specific gravity for flight, and providing room for the growth of internal organs. Usually, the first pair of spiracles is found on the mesothorax. Lepidoptera may control the flow by opening and closing the spiracles (**Figure 4**) [42–45].

1.6.4. Excretory system

The function of the excretory system is to maintain chemical homeostasis. By this system, hemolymph is cleaned with metabolic wastes including nitrogenous waste products created during digestion of food. As well as toxins, concentration of salts, and water are also regulated in hemolymph by the same. Malpighian tubules and the hindgut comprise the excretory system. Malpighian tubules, attached to the gut, float freely within the hemocoel and are bathed in hemolymph. They vary in number from 2 to 250 or more. Their functions are removing toxins, nitrogenous wastes, and ions to maintain ionic concentrations within the hemolymph. Water and other small ions are removed from the gut by the rectum (**Figure 4**) [42–50].

1.6.5. Nervous system

In Lepidoptera, the central nervous system is composed of a double chain of ganglia joined by longitudinal connectives. The anterior ganglion is the brain. The brain connects to the ventral chain of ganglia via two connectives that travel around the pharynx. The brain connects to the eyes, ocelli, and antennae. The subesophageal ganglion is highly complex and innervates the sense organs and muscles associated with the mouthparts, salivary glands, and neck region. The subesophageal ganglion is the primary excitatory or inhibitory influence on motor activity of the whole Lepidoptera. The frontal ganglion connects the brain to the stomatogastric subsystem. The hypocerebral ganglion is associated with two endocrine glands one of which is the corpus allatum that produces JH. The thoracic ganglia contain the sensory and motor centers for their respective segments. More derived taxa show a reduction in the number of abdominal ganglia. In visceral nervous system, nerves associated with the brain, salivary glands, and foregut are the stomatogastric subsystem. The caudal visceral subsystem is associated with the posterior segments of the abdomen including the reproductive system. In peripheral nervous system, all of the nerves are with synapses to the central and the visceral nervous systems. These nerves are associated with sensory structures (**Figure 4**) [51–59].

1.6.6. Reproduction

The adult male Lepidoptera reproductive tract is composed of a pair of testes, vas deferens, accessory glands, ejaculatory duct, and aedeagus. In testes, sperm begin to mature during the third and fourth larval instars. These divisions take place during the larval and pupal stages. Sperm are matured through spermiogenesis. All butterflies and moths produce two kinds of sperm: eupyrene sperm have a nucleus and can fertilize eggs, while apyrene sperm do not have a nucleus, and they facilitate the eupyrene sperm. Matured sperm are transferred within a protein-rich ejaculate called a spermatophore. It forms within the male's aedeagus and is transferred with sperm at the very end of copulation into the bursa copulatrix of female which can take up to 16 hours. The adult female Lepidoptera reproductive tract is composed of the bursa copulatrix, sperm duct, spermatheca, ovaries with ovarioles, and common oviduct. The end of the ovarioles is called the germarium, where oocytes are produced from the original germ cells. This process begins during the larval stage and continues in imago. Oocytes are covered with chorion, which forms in the last stage of oogenesis. However, male genitals include a valva, which is usually large, as it is used to grasp the female during mating. In female genitalia, there are three basic arrangements of openings for copulation, fertilization, and egg-laying. Firstly, in exoporian, an external opening that carries sperm from the copulatory opening of gonopore to the ovipore is found in Hepialidae and its related families. Secondly, in monotrysian, a single genital aperture near the end of the abdomen through which both copulation and egg-laying occur is found in primitive groups. Thirdly, in ditrysian, an internal duct that carries sperm with two distinct openings each for copulation and egg-laying is found in all the remaining groups (98%). As the egg passes down the common oviduct, few sperm are released from the spermatheca [33, 46]. Fertilization occurs just before an egg is about to be laid. High levels of JH circulating in adult butterflies cause eggs to mature in females and cause the male reproductive tract to develop. Diapause Lepidoptera, which reach sexual maturity after the overwintering period, have low levels of JH in their hemolymph (**Figure 5a–c**) [60–67].

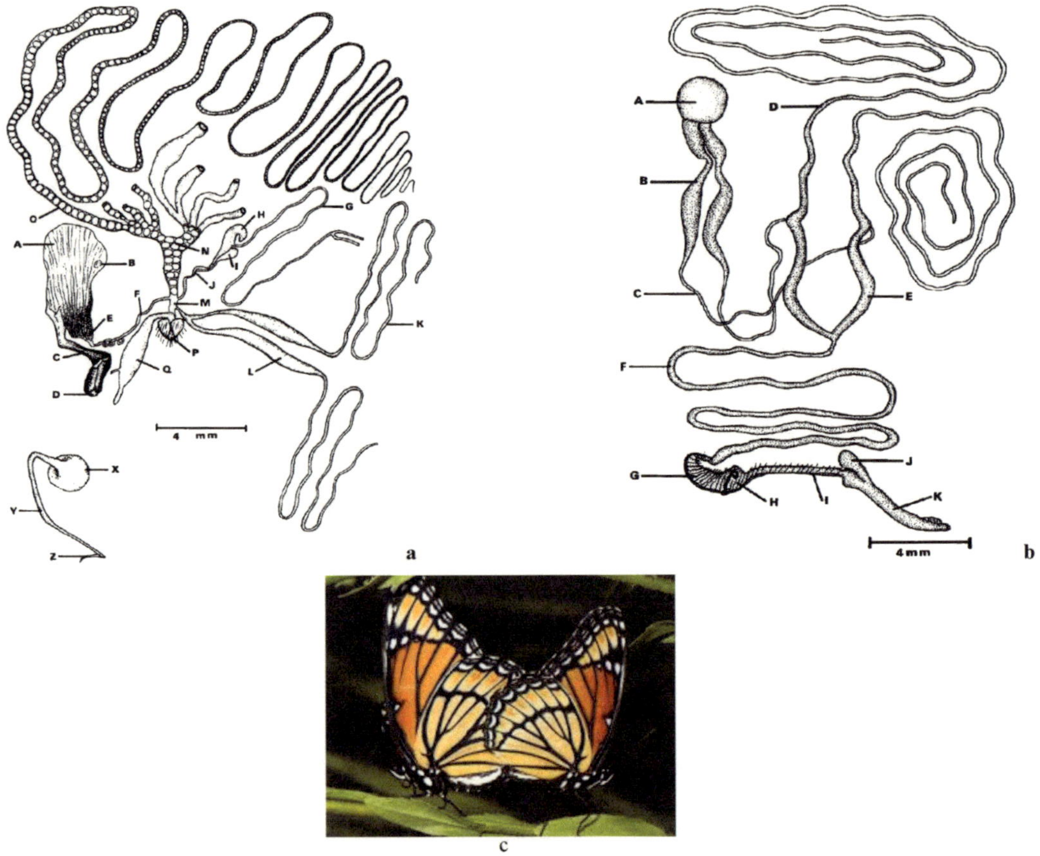

Figure 5. Anatomy of reproductive system of Lepidotera: (a) the female imago reproductive system: A: corpus bursae; B: signum, C: ductus bursae; D: ostium bursae; and E: diverticulum of bursa copulatrix; F: ductis seminalis; G: spermathecal gland; H: utriculus I: lagena of spermatheca; J: ductus receptaculi; K: accessory gland (paired); L: accessory gland reservoir (paired); M: vestibulum; N: calyx of the unpaired oviductus communis; O: one of four ovarioles of ovary (paired); P: papillae anales; Q: rectum; X: corpus, Y: collum, and Z: frenum of spermatophore; (b) the male imago reproductive system: A: testis; B: seminal vesicle (paired); C: vas deferens (paired); D: accessory glands (paired); E: ductus ejaculatorious duplex; F: primary segment of ductus ejaculatorious simplex; G: muascular area; H: area of frenum formation; and I: area of collum formation of the cuticular secondary segment of the ductus ejaculatorious simplex; J: caecum of aedeagus; K: aedeagus (After Etman and Hooper, 1979); (c) the mating pair of the the monarch butterfly, *Danaus plexippus* (Linnaeus, 1758) (Nymphalidae: Nymphalinae) [48, 49].

2. Lepidoptera as model taxon

Lepidoptera are ideal for to increase awareness toward environmental issues and educational purposes. They produce a more positive perspective of the invertebrates to the public, mostly due to their esthetic value. As Lepidoptera are the most well-known insects, they have become flagship organisms for the divulgation of invertebrate conservation plans. Their ecological significance is massive, not only because of the greatest percentage of species and biomass they account for in ecosystems, although they act as herbivores, pollinators, and food for insectivores. For researchers, scientists, and students, they offer a model taxon for precious to cram of biodiversity, conservation studies, environmental impact estimates, monitoring of animal populations, ecology, ethnology, evolution, genetics, systematic, and many other ecological and genetic studies. They open doors to establishment of chemical ecology as a scientific discipline for study and research.

Author details

Farzana Khan Perveen[1]* and Anzela Khan[2]

*Address all correspondence to: farzana_san@hotmail.com

1 Department of Zoology, Shaheed Benazir Bhutto University (SBBU), Main Campus, Sheringal, Khyber Pakhtunkhwa (KP), Pakistan

2 Roots Millennium College (RMC), Sector I-9, Islamabad, Pakistan

References

[1] Miller JC, Hammond PC. Lepidoptera of the Pacific Northwest Caterpillars and Adults. Forest Health Tecchnology Enterprise Team (FHTET), Technology Transfer. USA: USAD; 2003. p. 1-323

[2] Shield O. World number of butterflies. Journal of Lepidopterists Society. 1989;**43**:178-183

[3] Sutton P, Sutton C. How to Spot Butterflies. Boston, USA: Houghton Mifflin Company; 1999. p. 1-141

[4] Mallet J. Taxonomy of Lepidoptera: The Scale of the Problem. The Lepidoptera Taxome Project. London, UK: University College; 2007. p. 1-56

[5] Khan AG, Azim A, Nadeem MA, Ayaz M. The effect of formaldehyde treatment of solvent and mechanical extracted cottonseed meal on the performance, digestibility and nitrogen balance in lambs. Asian-Australian Journal Animal Science. 2000;**13**(6):785-790

[6] Khan ZR, Midega CAO, Wadhams LJ, Pickett JA, Mumuni A. Evaluation of Napier grass (*Pennisetum purpureum*) varieties for use as trap plants for the management of African stemborer (*Busseola fusca*) in a push-pull strategy. Entomologia Experimentalis Applicata. 2007;**124**(2):201-211

[7] Haroon K, Perveen F. Distribution of butterflies (family: Nymphalidae) in union council Koaz Bahram Dheri, Khyber Pakhtunkhwa, Pakistan. Social and Basic Sciences Research Review (SBSRR). 2015;**3**(1):52-57

[8] Perveen F, Khan A. Sikandar. Characteristics of butterfly (Lepidoptera) fauna from Kabal, swat, Pakistan. Journal of Entomology and Zoology Studies. 2014;**2**(1):56-69

[9] Danks HV. Insect dormancy: An ecological perspective. Biological Survey of Canada, Ottawa, Canada. 1987:1-32

[10] Fox LE, Soll DR, Wu CF. Coordination and modulation of locomotion pattern generators in drosophila larvae: Effects of altered biogenic amine levels by the tyramine beta hydroxylase mutation. Journal of Neuroscience. 2006;**26**(5):1486-1498

[11] Samways MJ. Insect Diversity Conservation. London, UK: Cambridge University Press; 1995. p. 1-342

[12] Boggs CL, Watt WB, Ehrlich PR. Butterflies: Ecology and Evolution Taking Flight. Chicago: University of Chicago Press; 2003. p. 1-345

[13] Hegna TA, Legg DA, Møller OS, Van RP, Rudy LA. The correct authorship of the taxon name Arthropoda. Arthropod Systematics and Phylogeny. 2013;**71**(2):71-74

[14] Scoble MJ. The Lepidoptera: Form, Function and Diversity. USA: Oxford University Press, Natural History Museum; 1992. p. 1-404

[15] Vukusic P. Structural color in Lepidoptera. Current Biology. 2006;**16**(16):R621-R623. DOI: 10.1016/j.cub.2006.07.040

[16] Krenn HW. Feeding mechanisms of adult Lepidoptera: Structure, function, and evolution of the mouthparts. Annual Review of Entomology. 2010;**55**:307-327. DOI: 10.1146/annurev-ento-112408-085338

[17] Hall JPW, Harvey DJ. A survey of androconial organs in the Riodinidae (Lepidoptera). Zoological Journal of the Linnean Society. 2002;**136**(2):171-197. DOI: 10.1046/j.1096-3642.2002.00003.x

[18] EBI, Encyclopedia Britannica, Inc., Oxford University Press, London, UK. 2014;**1**(32):1-640

[19] Perveen FK. Biology and Ecology of the Common Cutworm, *Spodoptera litura*. Raleigh, North Carolina, USA: Lulu Publishing and Marketing Inc., (LPM); 2016. p. 1-137

[20] Chapman RF. The Insects: Structure and Function (4 Ed.). New York, USA: Cambridge University Press; 1998. p. 1-715

[21] Bernhard CG, Ottoson D. Comparative studies on dark adaptation in the compound eyes of nocturnal and diurnal Lepidoptera. Journal of General Physiology. 1960;**44**(1):195-203

[22] Borror DJ, Triplehorn CA, Johnson NF. The Anatomy, Physiology, and Development of Insects: An Introduction to the Study of Insects. 6th ed. New York, USA: Saunders College Publishing; 1992. p. 24-73

[23] Jantzen B, Eisner T. Hindwings are unnecessary for flight but essential for execution of normal evasive flight in Lepidoptera. Proceedings of the National Academy of Sciences. 2008;**105**(43):16636-16640. DOI: 10.1073/pnas.0807223105

[24] Scoble M. The Lepidoptera: Form, Function and Diversity. London, UK: Oxford University Press; 1995. p. 66-67

[25] Borror DJ, Triplehorn CA, Johnson NF. An Introduction to the Study of Insects. 6th ed. Orlando, Florida. USA: Saunders College Publishing; 1992. p. 1-875

[26] Albrecht A, kaila L. Variation of wing venation in Elachistidae (Lepidoptera: Gelochioidea): Methodology and implications to systematic. Systematic Entomolgy. 1997;**22**:185-198

[27] Bashar MA, Mamun HRK, Chowdhury MAK. Wing-venation as a factor for the identification of Nymphalid butterflies in the forests of Chittagong and cox's bazar (Bangladesh). Bangladesh. Journal of Zoology. 2006;**34**(2):187-193

[28] Mathew G. Variation in the wing venations of *Pteroma Plagiophleps* Hampson (Lepidoptera: Psychidae). Journal of research on the Lepidoptera. 1985;**24**(4):359-336

[29] Mani MS. Ecology and Biogeography of High Altitude Insects. Vol. 4. New York, USA: Springer; 1968. p. 1-530

[30] Yack JE, Otero LD, Dawson JW, Surlykke A, Fullard JH. Sound production and hearing in the blue cracker butterfly, *Hamadryas feronia* (Lepidoptera: Nymphalidae) from Venezuela. Journal of Experimental Biology. 2000;**203**(24):3689-3702

[31] Lane KA, Lucas KM, Yack JE. Hearing in a diurnal, mute butterfly, *Morpho peleides* (Papilionoidea: Nymphalidae). Journal of Comparative Neurology. 2008;**508**(5):677-686. DOI: 10.1002/cne.21675

[32] Yack JE, Kalko EK, Surlykke A. Neuroethology of ultrasonic hearing in nocturnal butterflies (Hedyloidea). Journal of Comparative Physiology. 2007;**193**(6):577-590

[33] Gullan PJ, Cranston PS. The Insects: An Outline of Entomology. 3rd ed. Oxford, London: Blackwell Publishing; 2005. p. 22-48

[34] Singh UV, Awasthi VB. The neurosecretory cells and retrocerebral endocrine glands of Amsacta collaris Hampson (Lepidoptera: Arctiidae) during post-embryonic development. Journal Fur Hirnforschung 1981;**22**(3):265-278

[35] Patricia S, Clay S. How to Spot Butterflies. Boston, USA: Houghton Mifflin Company; 1999. p. 1-141

[36] Khan H, Perveen F. Distribution of butterflies (family: Nymphalidae) in union council Koaz Bahram Dheri, Khyber Pakhtunkhwa, Pakistan. Social and Basic Sciences Research Review (SBSRR). 2015;**3**(1):52-57

[37] Ando T, Inomata SI, Yamamoto M. The chemistry of pheromones and other semiochemicals. Topics in Current Chemistry. 2004;**239**:51-96

[38] Roelofs WL, Rooney AP. Molecular genetics and evolution of pheromone biosynthesis in Lepidoptera. Proceeding of National Academy of Science USA. 2003;**25:100**(Suppl 2):14599. DOI: 10.1073/pnas.1233767100a

[39] Sparks TH, Dennis RLH, Croxton PJ, Martin CADE. Increased migration of Lepidoptera linked to climate change. European Journal of Entomology. 2007;**104**(1):139-143

[40] Price M. Monarch miscalculation: Has a scientific error about the butterflies persisted for more than 40 years? Science News. 2017:1-1. DOI: 10.1126/science.aal0818

[41] Wikipedia. https://en.wikipedia.org/wiki/Lepidoptera#/media/File:Internal_morphology_of_Lepido ptera.svg. 2017; 1-2. (Retrieved: 05-05-2017)

[42] Capinera JL. Butterflies and moths. In: Encyclopedia of Entomology. 2nd ed. New York, USA: Springer; 2008; 4. p. 626-672

[43] Underwood DLA. Internal anatomy and physiology. Biology and general. Entomology. 2015;**1**:1-18

[44] Niels P, Kristensen L, Malcolm J, Scoble J, Karsholt O. Lepidoptera phylogeny and systematics: The state of inventorying moth and butterfly diversity. Zootaxa. 2007;**1668**:699-747

[45] de Gruyter. Hand Book of Zoology: Lepidoptera, Moths and Butterflies. Berlin, Germany: Verleg Publisher; 2003. p. 1-567

[46] Buntin GD, Pedigo LP. Morphology of the male and female reproductive systems of *Plathypena scabra* (F.) (Lepidoptera: Noctuidae). Journal of the Kansas Entomological Society. 1983;**56**(3):377-386

[47] Boggs CL. Reproductive strategies of female butterflies: Variation in and constraints on fecundity. Ecological Entomology. 1986;**11**:7-15

[48] Perveen F. Effects of Chlorfluazuron on the Reproduction of *Spodoptera litura*. In: Müller AG, editor. Berlin, Germany: VDM Verlag House Ltd.; 2009. p. 1-184

[49] Perveen F. Chlorfluazuron and embryonic development of *Spodoptera litura*. In: Müller AG, editor. VDM Verlag House Ltd., Berlin, Germany. 1-115.

[50] Covell CC Jr. A Field Guide to Moths. Boston,USA: Houghton Mifflin Co.; 1984. p. 1-496

[51] Covell CC Jr. The butterflies and moths of Kentucky: An annotated checklist. Kentucky State Nature Preserves Commission, Scientific and Technical Series Number. 1999;**6**:1-220

[52] Ferguson DC. Host records for Lepidoptera reared in eastern North America. US Department of Agriculture, Technical Bulletin. 1975;**1521**:1-49

[53] Ferguson DC, Chuck EH, Paul AO, Richard SP, Michael P, Jerry AP, Michael JS. Moths of North America, Jamestown, ND. Northern Prairie Wildlife Research Center Home. 1999:1-56

[54] Forbes WTM. The Lepidoptera of New York and neighboring states: Primitive forms, Microlepidoptera, Pyraloids, Bombyces. Cornell agricultural experimental state. Master of Engineering and Management. 1923;**193268**:1-729

[55] Forbes J. Lepidoptera of New York and neighboring states: Part II. *Geometridae, Notodontidae, Sphingidae,, Lymantriidae*. Cornell agricultural experimental state. Master of Engineering and Management. 1948;**274**:1-263

[56] Forbes J. Lepidoptera of New York and neighboring states: Part IV. Agaristidae through Nymphalidae. Cornell agricultural experimental state. Master of Engineering and Management. 1960;**371**:1-188

[57] Harp L, Chuck E, Paul AO, Richard SP, Michael P, Powell JA, Michael JS. Moths of North America. Jamestown, ND: Northern Prairie Wildlife Research Center; 2000. p. 1-9

[58] Hodges RW, et al. Checklist of the Lepidoptera of America North of Mexico. London: EW Classey Ltd.; 1983. p. 1-284

[59] Khan MR, Rafi MA, Nazir N, Khan MR, Khan IA, Hayat A, Ghaffar A, Rahim J, Perveen F. Biodiversity of butterflies from poonch division of Azad Kashmir, Pakistan. International journal of agricultural. Technology. 2014;**10**(4):885-898

[60] Perveen F, Khan A. Checklist of butterfly fauna from Kabal, swat, Pakistan. Journal of Advances in Biology. 2013;**02**(02):115-121

[61] Perveen F, Fazal F. Biology and distribution of butterfly fauna of Hazara university, garden campus, Mansehra, Pakistan. Special issue on Entomology Research: Open Journal of Animal Sciences (OJAS). 2013a;**3**(2A):28-36

[62] Perveen F, Fazal F. Checklist of butterfly fauna of Hazara university, garden campus, Mansehra, Pakistan. Signpost Open Access Journal (SOAJ) of Entomological Studies. 2013b;**2**(1):26-33

[63] Perveen F, Fazal F. Key for identification of butterflies (Lepidoptera) of Hazara university, garden campus, Mansehra, Pakistan. International Journal of Agriculture Innovations and Research. 2013c;**1**(5):156-161

[64] Perveen F, Ahmad A. Checklist of butterfly fauna of Kohat, Khyber Pakhtunkhwa, Pakistan. Art. 2012;**1**(3):112-117

[65] Perveen F, Ahmad A. Exploring butterfly fauna (Lepidoptera) of Kohat, Khyber Pakhtunkhwa, Pakistan. Signpost open access. Journal of Entomological Studies. 2012;**1**(2):94-107

[66] Perveen F. Distribution of butterflies (Lepidoptera) of Kohat, Khyber Pakhtunkhwa, Pakistan. Agricultural Science Research Journal. 2012;**2**(9):539-549

[67] Perveen F, Yasmin N, Ahmed A. Characterization of butterfly (Lepidoptera) of Kohat, Khyber Pakhtunkhwa, Pakistan. Pakistan Journal of Entomology, Karachi. 2012;**27**(1):15-26

Chapter 2

Lepidoptera Collection Curation and Data Management

Jurate De Prins

Additional information is available at the end of the chapter

http://dx.doi.org/10.5772/intechopen.70925

Abstract

The collections of Lepidoptera often serve as foundational basis for a wide range of biological, ecological, and climate science disciplines. Species identification and higher taxa delimitation based on collection specimens and especially, on types test scientific hypotheses, provide multiple types of evidence for a broad range of users. Curation and data management approaches applied in Lepidoptera collections benefit greatly from many newly developed information techniques, which link and integrate data. Mostly attention is focused on clean verified collection and taxonomic literature mining data to obtain correct species-group and higher taxa names, as well as reliable data on the distribution of Lepidoptera and their trophic interactions. Collection creation and management became a subject of natural sciences itself. The chapter provides a historic overview on collection creation and curation together with a short discussion on collection goals and purposes. The creation of a virtual collection based on interlinked data is emphasized. Information science and data management tools became very important in Lepidoptera collection curation. The complexity of techniques and computing tools used in taxonomy and the increase in the amount of data that can be obtained by collection-based disciplines make it necessary to automate data gathering, manipulation, analysis, and visualization processes.

Keywords: integrated collection, virtual collection, collection management tools, taxonomic text mining, data mining, web-based platforms, online catalogs

1. Introduction

The diversity of Lepidoptera is one of the most fascinating subjects of biology. Evolution, natural selection, and many other biotic and abiotic factors have produced different species of butterflies and moths and the speciation process is going on continuously. Present studies on Lepidoptera embrace many aspects on their function within the communities of plants and animals and a lot of different inter-relational processes that affect Lepidoptera. There are about 157,000 species of butterflies and moths currently described [1–3], in 135 families and 45 superfamilies [2]. Lepidoptera are a globally distributed, widely recognizable, and admired order of insects. Lepidoptera comprise ca. 10% of the total amount of described species of living organisms [3]. Lepidoptera are common in smaller or larger institutional collections and they are disproportionally abundant in private collections. Despite their popularity as one of the best known and most collected of all insect orders [4], there are no exact data available on how many species or specimens are deposited in natural history collections. At present, we know that about 17 million lepidopteran specimens are deposited in the collections of North America [5] and ca. 80% of all described Lepidoptera taxa are deposited in 60 European repositories [6], while the representatives of more than 38,000 species of moths described from the Afrotropical region are deposited in 158 natural history collections all over the world [7]. A rough estimation of the total Lepidoptera specimens in the depositories worldwide could be about 10% from the estimated 2.5 billion of natural history collection specimens [8–10].

Lepidoptera specimens in the collections of natural history document the present and historic delineation of species and higher taxa concepts which represent natural entities resulting from the differentiation of lineages through speciation with constantly changing boundaries [11]. In a collection, we deal with lepidopteran specimens sampled across vast geographical areas and through time [12, 13]. Another important aspect of the nomenclature of species is that it is based on type specimens, which means that any species name in Lepidoptera is eternally linked with the name-bearing type specimen, deposited in a public institutional collection which from the moment of publication serves as an unambiguous reference to the species name. Finally, collection is an endless source for large scale data:

i. taxonomic/nomenclatorial information related to the names of taxa;

ii. geographical data related to the distribution areas and biotopes;

iii. morphological data related to the delineation of species and higher taxa;

iv. biological/ecological data include valuable information on feeding and behavior habits of species within the complexity of interrelations; and

v. historical data related to the personalities of collectors and their activities.

Thus, the specimen is a natural history collection and its associated data serve as one of the most direct and reliable sources to answer numerous biodiversity research questions.

Novel technology adopted in Lepidoptera collections allows us to explore new horizons in the productivity of handling specimens, the curation of the collection and it significantly increases the quality of generated taxonomic data. The boom of digitization of natural history collections in recent years and the fast development of curatorial software allows easy access to the multiple collections of Lepidoptera spread over the world, and increases the use and reuse of valuable biodiversity data stored in those collections by providing access to species/specimen data through the Internet [8]. These data, ready to be incorporated into different models and virtual simulations, become crucial evidence for decision-making facing global problems such as climate change, species decline, habitat loss, pest monitoring, biological disaster predictions, and threats to agriculture and public health [14]. The major force driving the acceleration of interest and the use of data from the Lepidoptera collections is the unlimited digital access and interlinked visualized information. The usual practice of physical visits to a museum, negotiate with a curator for the access to the specimens, obtain permission following numerous internal regulations, and the financial and administrative restrictions related to them cause a serious bottleneck for collection-based research. Only a very limited number of people were privileged to have access to the valuable specimens and their associated data. This situation caused a huge taxonomic impediment, which means that despite the fact that collections contain a lot of novel data that need to be studied and incorporated into a broader pattern of the natural history data pool, these data were frozen in the collections for decennia [15–17]. In addition to this, collection curation became a dead-end professional career and a small group of people professionally engaged could not handle the broad scale of activities related to Lepidoptera biodiversity and collection data management in particular.

In this chapter, I intend to show that accessibility to a collection through the means of what new technology offers is the key to resolve a long-standing taxonomic impediment. A responsible and safe management of digitally interchangeable data provides new ways of handling different aspects and suggests new solutions for the complexity of problems related to biodiversity. Working with digital data, mined from the literature and Lepidoptera collections, is not the replacement of traditional methods by new ones, but rather it is the processing of the extracted data which have to pass the quality control by vetting and scrutinizing these data. It is the straight forward way to achieve what society needs at the moment: stable, long-lasting taxonomic decisions based on repeatable evidences influencing many aspects of society life.

2. Curation strategy of Lepidoptera collection

2.1. Historic approach

For centuries, Lepidoptera collections were created as part of curiosity objects, as a certain art of nature showing the interest of the owner to the world. The specimens were grouped in a certain order according to their size, geographic area, taxonomic knowledge of the owner, or other criteria. In the beginning of the twentieth century, many individual Lepidoptera collections moved as donations to museums or were purchased by public museums forming a major part of the holdings which the museums possess today. The role of a museum curator also developed in the course of time from the concept of ownership of collection cabinets (note: even the titles like Keeper (Natural History Museum, London) or Beheerder (Naturalis, Leiden)) indicate the attitude of possession, keeping, and administrating to the concept of a curator who collected and added specimens to the collection supporting taxonomic publications. The collections became reflections of the personal research of a scholar. Many expeditions were conducted for the need of finding new taxa for their taxonomic/curatorial research, based upon personal interests [18–20]. The concept of developing a Lepidoptera collection as a whole structural institutional unit reflects the result of historical taxonomic work done serving an integral and important part of the ongoing educational and research programmes of today, which was not developed at that time [6, 21]. There were no rules on priorities of collecting and processing the material. Curators were left to their own personal expertise. This collection keeping and managing style had a very huge negative effect on the next generation of taxonomists creating cross- and intra-institutional conflicts of interests. Further to this, many curators and researchers began to complain about the state of the current collections, the work involved to maintain and manage them [22–25].

At the same time, the curators became individuals collecting more and more specimens because biotopes and habitats were disappearing in an ever increasing speed. The collected specimens of numerous expeditions were stored without any processing and associated records. The primary core of a scholarly master of the collection disappeared. The taxonomic community at the end of the twentieth and the beginning of the twenty-first century is known as entering into the "crisis period" [26–28] and the collections seriously needed an effective "crisis management" strategy. Museum professionals started to regard the collections as a burden which consume finances and place, and not as strength of the museum. Because of this, the museums started hiring administrators and collection managers to better control the physical care over the collections. Individuals in these roles were preoccupied to create administrative rules and regulations, keeping track as where particular specimens were going. As a result, the clash between the collection curation and its administration increased.

2.2. Collection aims, objectives, and concepts

It is widely understood that habitat loss, land transformation, and habitat destruction are the major factors leading to the biodiversity loss. To understand how ecological systems change and interchange through space and time, reliable indicators are needed recording the state-of-the art of diversity, community composition within the framework of biotic and abiotic environmental conditions that facilitate these communities. Butterflies and moths possess multiple qualities that make them ideal as indicators. They are hyper-diverse, colorful, liked very much by many collectors of different ages, fill a wide range of functional roles like pollination, pest control, serving as prey in the complexity of food chains, nutrient cycles, have different population sizes, and life cycles, and respond rapidly to environmental changes [29, 30]. Lepidoptera communities in healthy habitats are often characterized by higher diversity, comprising a wide variety of taxa. Certain evolutional history or ecological aspects can be clarified by the presence or absence of specific taxa. For example, the presence of the micro-moth genus *Triberta*, De Prins et al. [31] might indicate the islands of a pre-glacial distribution pattern as well as the more recent colonization facilitated by human activities, since the genus is associated with the plant family Cistaceae [31]. Because of the overwhelming amount of information present already on Lepidoptera, the data must be sifted in order to identify those key species assemblages that can reveal the condition of the whole system and novel interactions.

The manual collection curation uses combined sources which are disparate and not always linked. At the heart of the collection curation, there are few but major ambitions and aims:

- to facilitate the access to the biodiversity resources and to present the physical illustration of the Lepidoptera biodiversity knowledge;

- to offer solutions to group and specify the biodiversity;
- to facilitate the direct and specific communication of collection users by providing a gate to the solid biodiversity platform related to all aspects of Lepidoptera;
- to provide inter-operability to data and resources at the highest level at the same time recognizing the fact that a well-curated and managed collection facilitates to further discoveries and application of novel methods in many areas of life sciences and disciplines of biological education.

Two domains, climate science and Big Data—which are directly connected to the Lepidoptera collection, are experiencing unprecedented attention, financing, and exponential growth. The curators of the collections address the Big Data challenges associated with climate science while incorporating the collection data into bigger data packages [14]. The main focus conceptually defining a collection is moved towards data analysis, because the taxonomic knowledge stored in a collection and gained from its interaction with other life and earth sciences produces biological Big Data that ultimately influence societal benefits [32]. The main concepts of the Lepidoptera collection can be defined as follows:

- national collection—reflecting the lepidopteran species of a certain country;
- continental or regional collection—reflecting the lepidopteran species of a certain continent or bio-geographical region;
- biological lepidopteran collection—reflecting species assemblages for agricultural or pest control purposes;
- taxonomic lepidopteran collection—reflecting the taxonomic accomplishment of a smaller or bigger taxonomic group, for example, family or superfamily;
- historic collection, which is usually obtained from the famous individual lepidopterist as his/her life achievement and reflects his/her personal views in defining the species concept within a particular group of Lepidoptera and at a certain epoch of time, for example, the Linnaeus collection.

It is important to acquire one of these abovementioned concepts of collection specialization and execute it creating a collection which becomes a tool-as-a-service within the framework of ongoing projects. Within a certain but defined framework of natural history problematics, a Lepidoptera collection plays an important role. Nevertheless, it is seen as a part of the constellation of sources which are a prerequisite to delivering the analytics to climate science as a service to society. Also a Lepidoptera collection serves as an educational tool, certain live textbook for a broad scale of people across many layers of society. Both elements of the Lepidoptera collection, (1) a tool-as-a-service and (2) a-tool-as-an-educational mean, are essential in handling and managing the Lepidoptera collection because in the aggregation with other domains of natural history, the data extracted from the collection lead to the generativity and assembly of interlinks which is the key of solving many of the Big Data challenges in the domains related to natural history. The creation of a Lepidoptera collection is an example of a building up verified data for a very long-term usage and enables a multi-sided retrospective analysis for research and applied lepidopterology.

Here below, I present a step-by-step methodology for an efficient, fast, reliable, easy searchable Lepidoptera collection curation. This approach can be easily applied and repeated by any mobile curatorial team in any museum of natural history housing Lepidoptera collections. The author has long-term experience curating large Lepidoptera collections of different ages and of different state in the biggest museums. The concept of a well-curated collection is that not only a curator but any authorized person easily finds any specimen he/she is looking for. The curated and completed collection expresses the concept adopted by a museum.

2.3. Delineating, identifying and describing the taxa

There are hot debates over the species concept and its biological reality [33]. While now there is a consensus to view species as natural entities delineated by multiple evidences resulting from differentiation of lineages through speciation [11], species boundaries are often much harder to discern especially in a Lepidoptera collection because specimens are sampled across vast geographical areas and through time [4, 11, 13, 18, 21]. Furthermore, the order Lepidoptera challenges species boundaries for regular potential or often occurring interbreeding [34]. Despite this, a species is a central concept in biology, conservation, legislation, and trade and therefore it has a particular social relevance to the museum collection and to the society.

The handling of specimens with great care and the associated meticulous documentation assist to the accurate identification of species. However, the contrasting phenomena of "over-splitting" or "over-lumping" of lepidopteran species take their turn due to the application of different methods and approaches in species delineation as well as biological and non-biological reasons [11], such as:

- similar populations which have been considered as distinct species;
- species complexes with little genetic information;
- ecological forms of polymorphic species (altitudinal, latitudinal, habitat, host-plant associations);
- variable species;
- incomplete lineage sorting, introgression, hybridization;
- inaccurate taxonomy, misidentifications, labeling errors, etc.

Though historic collections were created mainly based on the biological species concept, we, contemporary curators, are dealing with more and more consensual species concepts. Some genetic factors might play a major role in the species delineation of insects, such as intraspecific variation and the extent of divergence between species [11]. While curating a collection and identifying the species curators deal with complexes of monophyletic and/or non-monophyletic (polyphyletic and paraphyletic) species complexes which are very often difficult to distinguish [11]. Summarizing, it needs to be stressed that preference for accurate curation and attentive identification is one of the most important approaches in the delimitation of species or higher taxa. At present more and more research is focused on studies which extend beyond the taxonomic species descriptions, which crosses the taxa among bio-kingdoms and

http://dx.doi.org/10.5772/intechopen.70925

bio-classes emphasizing the trophic chains of relationships between plants, Lepidoptera, and other orders of insects [21]. Though taxonomy represents one of the most classical fields of life sciences, the new technology provides unlimited possibilities within this discipline to embrace novelties and to combine multiple evidences into a holistic pattern for the delineation of taxa within the order Lepidoptera.

2.4. Transfer from physical specimen to digital data collection

Some present academic educational studies and projects on tropical Lepidoptera often involve the creation of a physical collection [35] which is designed for the primary purpose:

i. to correctly identify species;

ii. to preserve specimens;

iii. to document the information related to specimens and species; and

iv. to make specimens available for scientific studies.

The focus is on having a vouchered collection which facilitates the precise identification of species (**Figure 1**). Only when parts of the collection are properly curated the specimens can be quickly and efficiently digitized. This approach makes the physical specimen collection a ready-to-use tool for any project requesting digitized data within a short period of time and leaves time and space for research related to other collection items. The collection transfer to twenty-first century systematics happens in two phases:

Stage 1. Having the matrix of major families curated and temporarily leaving families of insects in their old place in the collection.

Stage 2. Within 2–4 days transfer the families according to the modern systematics, which is now stabilized after the publications of high standard molecular papers.

The further steps are related to the computing of the collection data since information science plays a more and more important role in the collection curation and data management. This computed collection-based biodiversity data presentation aims to study, design, and develop solutions to automate the steps in the data gathering and data curation in

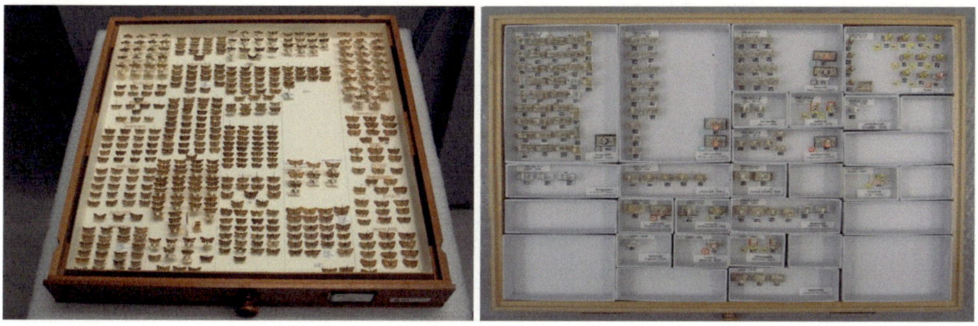

Figure 1. Historic collection and structured identified specimens collection.

order to reduce the drawbacks and difficulties in handling the huge data sets that have been provided for a study. Any curator/taxonomist/researcher, even without deep knowledge of computing, using the proper taxonomic tools is able to design and built his/her data matrices, gather the needed data and see the results in a comparable and convenient way (**Figure 2**).

A lot of surveys consist of cross-sectional studies and use a great amount of information gathered from the collection by means of different taxonomy-related queries. While the information obtained from these queries is very useful for both researchers and collection curation professionals, the management and analysis of data in many occasions is cumbersome and leads to a long process of search carried out by humans which also implies a possibility of errors. Data gathering for the collection is usually preceded by data mining of the literature sources. After that follows the process and visualization of data related to the collection using the appropriate photographing techniques and software designed specifically for the collection items. So, the professionals are responsible for obtaining the information in a format that is useful for their work. This process in the collection is the most time-consuming and error prone and might be subjectively biased. For some Lepidoptera collections, we have designed the tool (s) which have already passed the time and application test [36, 37] in three institutions: Royal Museum for Central Africa (Belgium), Natural History Museum (United Kingdom), and Royal Belgian Institute of Natural Sciences (Belgium).

Figure 2. Intelligent computerized curation of the microlepidoptera collection at the Natural History Museum, London.

The idea to use data mining tools in the collection-based research is getting the needed speed for many-sided approaches and studies: taxonomy, ecology, species interactions, biology, invasive patterns, host specifics or phylogenetic relationships. The obtained collection-based data assemblages demonstrate the complexity of biodiversity patterns. The further proper structuring of data by purposed collection data management tools enables to find the mechanisms which influence those particular biodiversity patterns as well as to obtain a clearer picture of the topic of ongoing research.

3. Digital data management of Lepidoptera collection

3.1. Creation of a virtual collection based on interlinked data

Collection digitization and putting the authorized data on the internet is a high priority at the moment since society not only needs but even demands to find authorized trustworthy data on the internet consultable at any moment and everywhere. The creation of a virtual collection has two main purposes:

- as a service to the community
- for data analysis for ongoing research.

There is no doubt that the future of collection management and collection consultation is digital. Much of society has already moved to the digital communication. Further development of a virtual collection of Lepidoptera has a strong emphasis on improving the existing digital collections and it is full heartily welcomed by the community of lepidopterists all over the world. Present achievement in informatics allows to operate the different aspects and relations concerning Big Data, and this is exactly what the Lepidoptera collection can provide for society and for any user beyond taxonomists. However, in order to succeed in creating a virtual collection, the steps taken should consequently follow a strict order and be completed:

Step 1. Data are **structured**, so they can be exported as Excel sheets and incorporated in any database.

Step 2. Data are presented in the same way **consequently**.

Step 3. Every data unit (species/specimen) should obtain a **unique code** and/or identifier, so the data are machine readable and operational.

Step 4. The biodiversity information is **intelligently text and data mined** from taxonomic literature and collection, intelligently verified and filtered.

Step 5. **Visualization** of a collection is a very important aspect, since it touches all layers and all groups of society. Furthermore, the visualized data becomes understandable for any user worldwide.

Step 6. Data are combined in a network of intelligent **relationships**. It not only enables to combine data in different groups and find correlations but also it saves a lot of time because any data are entered only once and the predefined reports are obtainable within seconds.

Data related to a Lepidoptera collection are arranged into different information packages: datasets (**Figure 3**) which are interlinked and integrated. However, at the same time these different datasets are easily independently consulted, independently displayed and new information can be independently added. These interlinked and integrated, but nevertheless independently operational datasets, serve as separate work packages for the simplified extraction of complex data. The protection of sensitive information within the interlinked work packages is foreseen also, since certain data and data packages can be made seen by authorized users only. I suggest to continue a well-tested formula of five interlinked and related data packages (**Figure 4**).

In this way it is easy to obtain clean data, to find trends in the present information packages, to keep track on specimens (types and vouchers) and to link taxonomic, morphological and DNA-related information to a concrete taxon (**Figure 5**): (species → subgenus → genus → tribus → subfamily → family → order).

The visualized dataset on Lepidoptera include computer-assisted automatic distribution atlases and unlimited possibilities in the presentation of image galleries, both types and verified voucher specimens.

Let me briefly mention the aspects of a database of Lepidoptera as a tool. There is a certain reluctance towards the databases in the community of curators, since the databasing is seen as an administratively imposed activity which takes a lot of time and gives little in return.

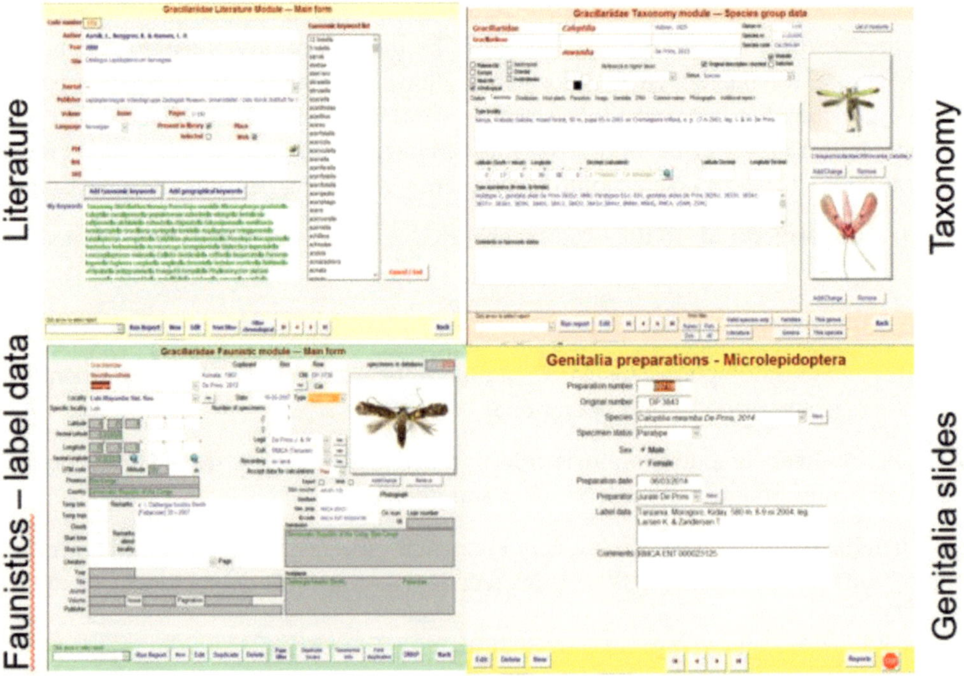

Figure 3. Simplified data entry behind the complex architecture.

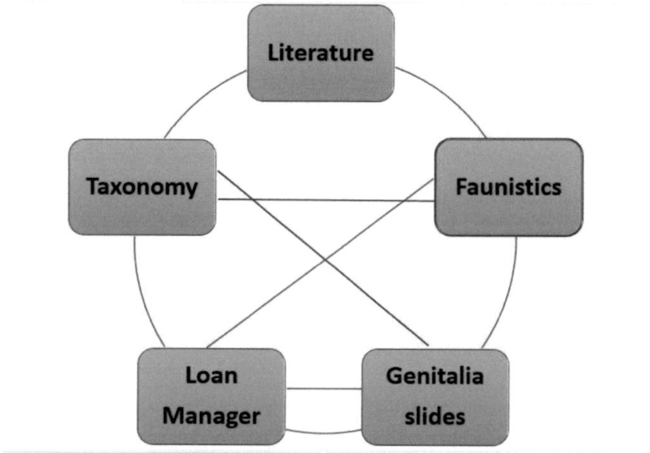

Figure 4. The structure of data: five interrelated and integrated datasets (following De Prins [36]).

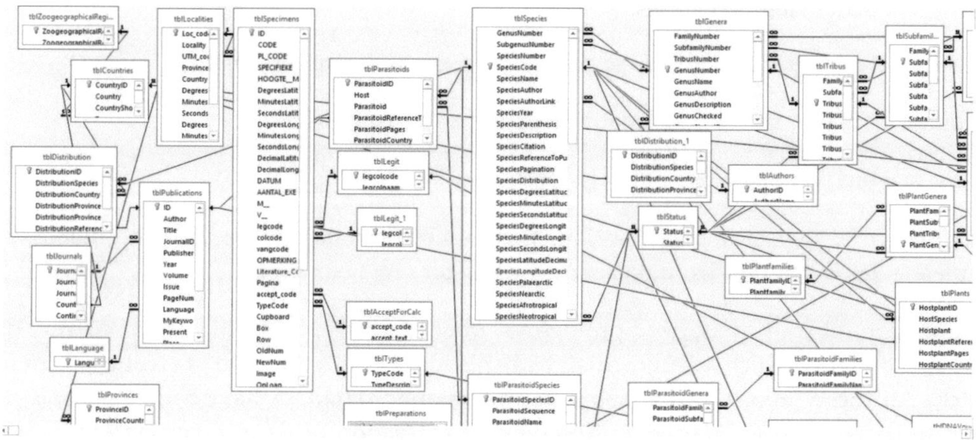

Figure 5. Part of the relationships of the dataset system (following De Prins [36]).

However, almost all curators-taxonomists working on a taxonomic group work with databases, because the information they have should be stored, accumulated, and consulted. What a good relational database can do for a curator:

- provide literature on taxa with reference to the exact pages of the original description and subsequent re-descriptions with indicated illustrations organized according to subject, in chronological or alphabetical order;
- provide correct taxonomic names, authors, dates;
- provide full and complete lists of synonyms with indicated sources;
- provide lists of taxa according to taxonomic classification or alphabetical order to arrange them in the collection;

- provide labels of taxa without typing errors or taxonomic mistakes;
- automatically make robust taxonomic catalogs and checklists and relates different biological data with references even with indicated pages;
- immediately assist locating the searched specimen in the collection, or if it is on loan to provide the details of the loan, so the curator always knows where any specimens under his/her care are located;
- show the type locality, even on Google Maps;
- show images of species, all their stages, host plants, habitats;
- provide data on genitalia or other micro-morphological structures, also images;
- provide lists of specimens belonging to the same species no matter where they are deposited;
- make a loan in a few minutes;
- show related species and help to make diagnoses and comparisons;
- quickly indicate the best time for organizing an expedition;
- show which types are deposited in which museum.

There can be a very strong motivation to compile the database because the outcomes of complete, related and verified data are rewarding for any taxonomist.

There are a number of specialized pre-defined queries and reports which are used very often in cyber cataloging which are immediately displayed without the need of creating them from scratch. All these reports in the suggested taxonomic database have a defined and fixed structure based on quantitative and qualitative extraction of data, mapping the data and interpreting these data based on the visual display of numerical charts. So both processes the curation of physical items (specimens) of the Lepidoptera collection and the integration of data into the predefined data matrices go together and can be largely automatized using unique digital ID scan-readable identifiers, predefined labels, loan forms, automatic monitoring of loans, automatic recognition of novelty in biological and distributional data, personalized data for collectors and donors, etc. This changes the way how Lepidoptera curators work in the collection and facilitates the process in order to eliminate errors due to human factors. When including societal and social media (e.g. www.waarnemingen.be or Facebook, Twitter, Instagram, ResearchGate) into the analysis of taxonomic data all the considerations presented above require greater relevance. Obtaining data on relationships in Lepidoptera from multiple digital communication records require to use the complex matrix and arrange data into smaller data packages (**Figure 4**) because the interrelated data is difficult to handle manually. The proper representation of basic faunistic data and associated species ID is crucial because this kind of information forms the basis for the later phase of analysis and visualization.

3.2. Online searchable catalogs

Many studies on taxonomy, as well as phylogenies of higher Lepidoptera groups are hampered by the fact that the world fauna of Lepidoptera is still not inventoried. Moreover, the current research tries to study the wider range of factors that may be involved into the evolutionary processes of lepidopteran species. In particular many different environmental factors in which species are immersed are of special interest in present approaches. The online cataloging aims to fill at least partly the gaps and provides the following primary data:

- the diversity of species;
- the spatial distribution pattern;
- the taxonomic and phylogenetic distinctiveness;
- nomenclature and study of primary types;
- synonymy;
- concise records of natural history;
- concise records of taxonomic history; and
- DNA accession numbers.

We presented two online catalogs: the catalog of one family of moths on a global scale available from www.gracillariidae.net and the catalog of all species of moths from a defined bio-geographical region, in our case the Afrotropics, available from www.afromoths.net. For the creation of searchable taxonomic online catalogs we used two modules of the dataset: (1) taxonomic and (2) literature of the interlinked dataset (**Figure 3**). The online searchable catalogs provide the referenced taxonomic and life history information in the following fields:

- family;
- subfamily;
- checked and correct species name;
- status (species, subspecies, synonym, unavailable name etc.);
- author;
- description year;
- original combination;
- Google mapped type locality;
- type specimens, associated genitalia slides and depository;
- publication of original description and pagination;
- distribution per country;
- biological data.

An approach based on the analysis of taxonomic data along with modern imaging techniques may give more insight into the taxonomic situation and group relationships than narrower, more specialized traditional studies. Nevertheless, the widely adopted approaches include a number of well-known, standardized data packages. They have been included into the proposed digital tools for online cataloging:

i. **taxonomy** (taxonomic position, current species name, synonymy, original combination and the reference to the original description with indicated pagination);

ii. **types** (name bearing type specimen(s): holotype, syntypes, lectotypes, neotype), other verified type specimens (paratypes, paralectotypes) and other not defined by Code (ICZN 1999) verified type specimens with associated institutional numbers, associated microscopic preparation slides, their depository place;

iii. **distribution** and **habitats** (mapped and referenced distributional data);

iv. **biology** (referenced data on food plants, and concise life history data);

v. **DNA** (accession numbers of the GenBank and linked to them the DNA information).

These interlinked data packages serve to obtain qualitative online catalogs that present data on Lepidoptera biodiversity displayed into a number of categories. Online cataloging standardizes taxonomy all over the world and makes the collection curation a fast and finalized endeavor in any museum at any place in the world. It becomes also possible to combine data obtained from many different museums into one huge data network and to fill the gaps in taxonomy and other related disciplines.

3.3. Data management and analysis

The proposed approach of data management [36, 37] is based on possibilities that any professional researcher or citizen scientist retrieves personalized taxonomic, trophic, and distribution data needed for research or study. The proposed architecture of taxonomic data management has a web application function and a robust image gallery. On server side it has been developed with the assistance of GBIF, BeBIF, Catalog of Life as the up-to-date database management system. The literature records have been used for displaying the referenced data which in many cases are also linked with the Biodiversity Heritage Library.

Four different user roles can be defined in the taxonomic data management system:

i. Supervisor Administrator who has full permissions to manage and manipulate data, to create queries and reports, answer any taxonomic question or produce structured data sets in an exportable format (excluding access to data described in data protection laws).

ii. Taxonomist-Administrator who validates the taxonomic information, nomenclatural issues, homonymy, synonymy, availability of names and links with the checked reference; produces robust global or regional interlinked taxonomic catalogs.

iii. Taxonomist-Identifier who adds associated images to species pages, host plant(s), and distribution information.

iv. Collection Administrator who manages collection specimen data and administrates loans.

The display of results through the data management system eases the work of curators and presents the data in the way the user needs:

Standardized data management. The user obtains all the data in a strictly standardized format and can deeply inspect them as well as the general information meaningful to the search question.

Customized data management. The user can design and create the combinations of data of interest. Also data can be grouped into sets based on predefined formula in the similar way to the validated queries. The data management system will automatically perform the display of data from the requested search fields.

Visualized survey of data. This part of data display shows the imaged specimens once they are identified and curated. As well as the plain qualitative and quantitative taxonomic results obtained from data curation and data management there is part of data presentation which is visualized. This type of information showing shapes, patterns and colors of Lepidoptera and representing different characteristics of taxa/specimens has been carried out by the application of micro/macro photography techniques. Digital imaging analysis tools for Lepidoptera have been extensively used for a number of studies and websites during the first decade of this century. New imaging tools exist today that have been designed for collection specimens with capacities to overcome the shortcomings of traditional micro/macro photography and are applicable for mobile devices. In addition to microphotography, the microtomography has been proven as a promising technique to allow the visualization of internal morphological structures in non-destructive way and present them three-dimensionally.

3.4. Intelligently assembled and curated collection

The internet-linked data analysis of expert assembled and curated Lepidoptera collection integrates observational data into many possible models. An intelligently assembled and curated Lepidoptera collection represents a data product that is of growing importance to researchers working in the domain of climate science and preoccupied with a wide range of applications which need fast decision procedure. The Lepidoptera collection, as a more or less completed institutional unit, brings together the following set of elements:

1. high reliability and performance of data analysis;
2. data management on smaller or bigger scale;
3. appliance of virtualization and user attractive visualization;
4. possibilities to adaptive usage of data;
5. harmonization of data within the domain of natural sciences.

The effectiveness of the internet-linked data extracted from the Lepidoptera collections has been demonstrated in several successful case studies [7, 8, 12, 38]. Structuring and digitization of collection data and presenting them in an internet-linked environment lowers the barriers for obtaining the taxonomic scholarship, democratizes the taxonomic community of lepidopterists, fosters innovation and experimentation with the collection data, facilitates the usage of technology with the collection items, and provides the agility required to meet the multi-purpose needs of users. The structured, internet-linked taxonomic and collection-based data are providing new data service within natural history that helps to connect academic and computational resources. Moreover, the structured, internet-linked Lepidoptera collection data engage the multinational communities of naturalists and climate science specialists in the construction of new capabilities. The provision of such interchangeable multi-purpose data probably is one of the most important changes in the way we, museum-based curators and taxonomists work within the modern society.

New technological and societal developments shifted significantly the paradigm what collection curation and what curators are. For 200 years lepidopteran taxonomists followed individual strategies. Later the taxonomy of Lepidoptera saw the inclusion of molecular approaches and techniques. Despite these new technologies, the integrative delimitation of lower taxa at the genus and species level proceeds to be a continuously ongoing process parallel with the descriptions of novel taxa and requires a taxonomic consensus which actually needs to be included into the data matrices on a daily basis. Also, an intelligently curated collection of Lepidoptera is a live tool changing daily due to changes in taxonomy or additions in biogeography. Taxonomists/curators may feel uneasy with increasingly time consuming and laborious work while delineating taxa, but also scientists find it difficult for obtaining comparative data if the reference collection is not curated or is curated poorly.

A curated and managed high quality collection will overcome the problems that hamper data search and interoperability between taxonomic and research labs. The use of incorrect and invalid names will result in heterogeneous, incomplete, fragmented datasets which need verification [39]. The application of a unique numbering system for taxa facilitates machine-readable linkages to data sources, easily detects any human-caused error [36, 38] and speeds up the connectivity in ever expanding mass of taxa and data affiliated to them. Digital collection curation is fully in line with ideas to link and share collection-based data and to explore all resources available in public institutions and private holdings. Collection curation based on a long tradition as a non-profit occupation allows and facilitates public access to lepidopteran biodiversity worldwide and adds a needed value to the already known and still unknown lepidopteran biodiversity in such a way that a Lepidoptera collection becomes a ready-to-use tool for science, scientists, and society. The collection allows us to exploit the sources and knowledge from different aspects, to discover and disclose new findings within the framework of global science of natural history.

4. Conclusions

Information science and data management tools have become very important in the curation of Lepidoptera collections. The complexity of techniques and computing tools used in taxonomy and the increase in the amount of data that can be obtained from collection-based disciplines make it necessary to automate processes in data gathering, manipulation, analysis and visualization. Much data used in taxonomy and Lepidoptera collection management comes from unverified offline taxonomic datasets and specimen labels. This can lead to time-consuming and error-prone processes that can be easily automated. In this sense, the collaboration between researchers, taxonomists, citizen scientists, collection curators, and computing/information science is crucial to build and to use the proper approaches in taxonomy needed to avoid error-prone situations and to obtain qualitative results without the need of being experts in a certain taxonomic group or in the techniques underlying the automated processes. Modern approaches towards Lepidoptera collections and data management help to focus on the goals and studies that can be finalized.

For future work, I see a much closer integration of different disciplines related to life and climate sciences and inclusion of new functionalities into the offline and online tools that could provide much deeper insights into the diversity of Lepidoptera as well as into the complexity of relationships, thus improving the usefulness of these tools for research and identification purposes. The structured, searchable global, and regional databases of Lepidoptera have already been of significant assistance in the evaluation of Lepidoptera diversity at national and international levels and in the curation of large institutional collections. The novel approaches in curation, data management, and collection-based science can also be incorporated into educational programs so that the lepidopterist community and society in general can test, use and explore all possible benefits from Lepidoptera collections.

Acknowledgements

I thank my lepidopterists colleagues and curators of Lepidoptera collections for valuable discussions during many years of collaboration. My special thanks are extended to Willy De Prins, Stefan Kerkhof, and Wouter Dekoninck (Royal Belgian Institute of Natural Sciences, Brussels, Belgium) for their inspiring talks and strong support for Lepidoptera collection-based issues. The IT team of the Belgian Biodiversity Platform and the Global Biodiversity Information Facilities is thanked for their technical assistance in launching the structured taxonomic data on Lepidoptera online. Donald and Mignon Davis (Smithsonian Institution, Washington DC, USA) are cordially acknowledged for their valuable comments.

Author details

Jurate De Prins

Address all correspondence to: jurate.deprins@gmail.com

Royal Belgian Institute of Natural Sciences, Brussels, Belgium

References

[1] Kristensen NP, editor. Lepidoptera, Moths and Butterflies. Volume 1: Evolution, Systematics, and Biogeography. Handbook of Zoology IV (35). Berlin, New York: Walter de Gruyter; 1998. 487 p

[2] van Nieukerken EJ, Kaila L, Kitching IJ, Kristensen NP, Lees DC, Minet J, Mitter C, Mutanen M, Regier JC, Simonsen TJ, Wahlberg N, Yen S-H, Zahiri R, Adamski D, Baixeras J, Bartsch D, Bengtsson BÅ, Brown JW, Rae Bucheli S, Davis DR, De Prins J, De Prins W, Epstein ME, Gentili-Poole P, Gielis C, Hättenschwiler P, Hausmann A, Holloway JD, Kallies A, Karsholt O, Kawahara AY, Koster S(JC), Kozlov MV, Lafontaine JD, Lamas G, Landry J-F, Lee S, Nuss M, Park K-T, Penz C, Rota J, Schintlmeister A, Schmidt BC, Sohn J-C, Solis MA, Tarmann GM, Warren AD, Weller S, Yakovlev RV, Zolotuhin VV, Zwick A. Order Lepidoptera Linnaeus, 1758. In: Zhang Z-Q, editor. Animal biodiversity: An outline of higher-level classification and survey of taxonomic richness. Zootaxa. 2011;**3148**:212-221

[3] Mallet J. The Lepidoptera Taxome Project [Internet]. 2014. Available from: http://www.ucl.ac.uk/taxome/ [Accessed: 2017-07-17]

[4] Kawahara AY, Pyle RM. An appreciation for the natural world through collecting, owning and observing insects. In: Lemelin RH, editor. The Management of Insects in Recreation and Tourism. Cambridge University Press; 2012. p. 138-152

[5] Seltmann KC, Cobb N, Gall LF, Barlett CR, Basham A, Betancourts I, Bills C, Brandt B, Brown RL, Bundy C, Caterino M, Chapman C, Cognato A, Colby J, Cook SP, Daly KM, Dyer LA, Franz NM, Gelhaus JK, Grinter CC, Harp CE, Hawkins RL, Heydon SL, Hill GM, Huber SH, Johnson N, Kawahara AY, Kimsey LS, Kondratieff BC, Krell F-T, Leblanc L, Lee S, Marshall CJ, McCabe LM, McHugh JV, Menard KL, Opler PA, Palffy-Muhoray N, Pardikes N, Peterson M, Pierce NE, Poremski A, Sikes DS, Weintraub JD, Wikle D, Zaspel J, Zolnerowich G. LepNet: The Lepidoptera of North America network. Zootaxa. 2017; **4247**:73-77. DOI: 10.11646/zootaxa.4247.1.10

[6] CETAF–Consortium of European Taxonomic Facilities. Available from: http://cetaf.org [Accessed: 2017-07-17]

[7] De Prins J, De Prins W. AfroMoths, an Online Database of Afrotropical Moth Species (Lepidoptera). 2017. Available from: http://www.afromoths.net/ [Accessed: 2017-07-17]

[8] Güntsch A, Hyam R, Hagedorn G, Chagnoux S, Röpert D, Casino A, Droege G, Glöckler F, Gödderz K, Groom Q, Hoffmann J, Holleman A, Kempa M, Koivula H, Marhold H, Nicolson N, Smith VS, Triebel D. Actionable, long-term stable and semantic web compatible identifiers for access to biological collection objects. Database – The Journal of Biological Databases and Curation. 2017;**2017**:1-9. DOI: 10.1093/database/bax003

[9] Duckworth WD, Genoways HH, Rose CL. Preserving Natural Science Collections: Chronicle of Our Environmental Heritage. Washington, DC: National Institute for the Conservation of Cultural Property; 1993. 140 p

[10] Nudds JR, Pettitt CW, editors. The Value and Valuation of Natural Science Collections. London: The Geological Society; 1997. 240 p

[11] Mutanen M, Kivelä SM, Vos RA, Doorenweerd C, Ratnasingham S, Hausmann A, Huemer P, Dincă V, van Nieukerken EJ, Lopez-Vaamonde C, Vila R, Aarvik L, Decaëns T, Efetov KA, Hebert PDN, Johnsen A, Karsholt O, Pentinsaari M, Rougerie R, Segerer A, Tarmann G, Zahiri R, Godfray HCJ. Species level para- and polyphyly in DNA barcode gene trees: Strong operational bias in European Lepidoptera. Systematic Biology. 2016;**65**:1024-1040. DOI: 10.1093/sysbio/syw044

[12] Chapman AD. Principles of Data Quality, Version 1.0. Report for the Global Biodiversity Information Facility: Copenhagen; 2005

[13] Hofmann AF, Tremewan WG. The Natural History of Burnet Moths. Museum Witt Munich & Nature Research Center Vilnius: Vilnius; 2017. 630 p

[14] Schnase JL, Duffy DQ, Tamkin GS, Nadeau D, Thompson JH, Grieg CM, McInermey MA, Webster W. MERRA analytic services: Meeting the big data challenges of climate science through cloud-enabled climate analytics-as-a-service. Computers, Environment and Urban Systems. 2017;**61**:198-211. DOI: 10.1016/j.compenvurbsys.2013.12.003

[15] Hebert PDN, deWaard JR, Zakharov EV, Prosser SWJ, Sones JE, McKeown JTA, Mantle B, La Salle J. A DNA 'barcode blitz': Rapid digitization and sequencing of a natural history collection. PLoS One. 2013;**8**:e68535. DOI: 10.1371/journal.pone.0068535

[16] Cavallin EKS, Munhoz CBR, Harris SA, Villarroel D, Proença CEB. Influence of biological and social-historical variables on the time taken to describe an angiosperm. American Journal of Botany. 2016;**103**:2000-2012. DOI: 10.3732/ajb.1600120

[17] Pons J, Barraclough TG, Gomez-Zurita J, Cardoso A, Duran DP, Hazell S, Kamoun S, Sumlin WD, Vogler AP, Hedin M. Sequence-based species delimitation for the DNA taxonomy of underscribed insects. Systematic Biology. 2006;**55**:595-609. DOI: 10.1080/10635150600852011

[18] Epstein ME. Moths, Myths & Mosquitoes. The Eccentric Life of Harrison G. Dyar Jr. New York: Oxford University Press; 2016. 325 p

[19] Clarke JFG. Catalogue of the Type Specimens of Microlepidoptera in the British Museum (Natural History) Described by Edward Meyrick. London: British Museum; 1955. 332 p

[20] Gilbert P. A Source Book for Biographical Literature on Entomologists. Leiden: Backhuys Publishers; 2007. 694 p

[21] Strategy to 2020. Advancing the Science of Nature. London: The Natural History Museum; 2015. 22 p

[22] Suarez AV, Tsutsui ND. The value of museum collections for research and society. Bioscience. 2004;**54**:66-74. DOI: 10.1641/0006-3568(2004)054[0066:TVOMCF]2.0.CO;2

[23] Ponder WF, Carter GA, Flemons P, Chapman RR. Evaluation of museum collection data for use in biodiversity assessment. Conservation Biology. 2001;**15**:648-657

[24] Holmes MW, Hammond TT, Wogan GOU, Walsh RE, LaBarbera K, Wommack EA, Martins FM, Crawford JC, Mack KL, Bloch LM, Nachman MW. Natural history collections as windows on evolutionary processes. Molecular Ecology. 2016;**25**:864-881. DOI: 10.1111/mec.13529

[25] Cho S, Epstein SW, Mitter K, Hamilton CA, Plotkin D, Mitter C, Kawahara AY. Preserving and vouchering butterflies and moths for large-scale museum-based molecular research. PeerJ. 2016;**4**:e2160. DOI: 10.7717/peerj.2160

[26] Evans Walter D, Winterton S. Keys and the crisis in taxonomy: Extinction or reinvention? Annual Review of Entomology. 2007;**52**:193-208. DOI: 10.1146/annurev.ento.51.110104.151054

[27] Mallet J, Willmott K. Taxonomy: Renaissance or tower of babel? Trends in Ecology and Evolution. 2003;**18**:57-59

[28] Agnarsson I, Kuntner M, Paterson A. Taxonomy in a changing world: Seeking solutions for a science in crisis. Systematic Biology. 2007;**56**:531-539. DOI: 10.1080/10635150701424546

[29] Walther G-R, Post E, Convey P, Menzel A, Parmesan C, Beebee TJC, Fromentin J-M, Hoegh-Guldberg O, Bairlein F. Ecological responses to recent climate change. Nature. 2002;**416**:389-395

[30] Parmesan C. Ecological and evolutionary responses to recent climate change. The Annual Review of Ecology, Evolution and Systematics. 2006;**37**:637-669. DOI: 10.1146/annurev.ecolsys.37.091305.110100

[31] De Prins J, Davis D, De Coninck E, Sohn J-C, Triberti P. Systematics, phylogeny and biology of a new genus of Lithocolletinae (Lepidoptera: Gracillariidae) associated with Cistaceae. Zootaxa. 2013;**3741**:201-227. DOI: 10.11646/zootaxa.3741.2.1

[32] Ford JD, Tilleard SE, Berrang-Ford L, Araos M, Biesbroek R, Lesnikowski AC, MacDonald GK, Hsu A, Chen C, Bizikova L. Opinion: Big data has big potential for applications to climate change adaptation. PNAS. 2016;**113**:10729-10732. DOI: 10.1073/pnas.1614023113

[33] De Queiroz K. Species concepts and species delimitation. Systematic Biology. 2007;**56**: 879-886. DOI: 10.1080/10635150701701083

[34] Zhang L, Reed RD. A practical guide to CRISPR/Cas9 genome editing in Lepidoptera. bioRxiv preprint first posted online 2017-04-24. DOI: 10.1101/130344

[35] Brito R, De Prins J, De Prins W, Mielke OHH, Gonçalves GL, Moreira GRP. Extant diversity and estimated number of Gracillariidae (Lepidoptera) species yet to be discovered in the Neotropical region. Revista Brasileira de Entomologia. 2016;**60**:275-283. DOI: 10.1016/j.rbe.2016.06.002

[36] De Prins J. An integrated taxonomic tool for online dissemination of concise, verified and visualized information on biodiversity, retrieved from data and text mining of natural history collections and libraries. JSM Bioinformatics, Genomics and Proteomics. 2016;**1**: 1006

[37] De Prins J. Filling the gaps on trophic interactions between Lepidoptera and hymenoptera using information retrieved from interlinked and integrated taxonomic datasets. JSM Anatomy and Physiology. 2017;**2**:1006

[38] GBIF–Global Biodiversity Information Facility. Available from: http://www.gbif.org/ [Accessed: 2017-07-18]

[39] de Jong R. Fossil butterflies, calibration points and the molecular clock (Lepidoptera: Papilionoidea). Zootaxa. 2017;**4270**:001-063. DOI: 10.11646/zootaxa.4270.1.1

Chapter 3

Molecular Phylogeny and Taxonomy of *Lepidoptera* with Special Reference to Influence of *Wolbachia* Infection in the Genus *Polytremis*

Weibin Jiang

Additional information is available at the end of the chapter

http://dx.doi.org/10.5772/intechopen.69958

Abstract

This chapter provides a case of genus *Polytremis* Mabille, 1904 (*Lepidoptera*: Hesperiidae), to explain the molecular phylogeny and taxonomy of *Lepidoptera* and the influence of *Wolbachia* infection. Earlier studies of *Lepidoptera* were focused mainly on the morphological classification, population distribution, and identification of new species. As the supplementary to morphological research, analysis of DNA has been widely used in the phylogenetic studies of *Lepidoptera*. The study provides a conservative estimate that the *Wolbachia* infection rate in *Polytremis nascens* Leech (1893) is 31%, and no significant difference in the prevalence is found between the sexes. The *Wolbachia* infection mainly prevails in populations of *P. nascens* in southern China, which influence the diversity of mtDNA in *P. nascens* by a *Wolbachia*-induced sweep. The *Wolbachia* infection rate in *Polytremis fukia* Evans (1940) is 47% and shows a weak association existed between mitochondrial DNA haplotypes and wFuk1 infection status.

Keywords: *Lepidoptera*, microsatellite, mitochondrial genome, molecular phylogeny, taxonomy, *Wolbachia*

1. Introduction to molecular phylogeny and taxonomy of *Lepidoptera*

Butterflies and moths (*Lepidoptera*) have long served as a model system for ecological and evolutionary studies on the basis of the high degree of diversity and complexity, which constitute one of the most diverse insect orders with more than 157,000 described species. Earlier studies of *Lepidoptera* were focused mainly on the morphological classification, population distribution, and identification of new species. As the supplementary to morphological research, analysis of

DNA has been widely used in the phylogenetic and taxonomic studies of *Lepidoptera*. The case of genus *Polytremis* will be discussed as follows.

2. The phylogeny of the butterfly genus *Polytremis*

The family Hesperiidae includes more than 4000 species, commonly known as "skippers," of which Hesperiinae is the largest subfamily. *Polytremis* Mabille (1904) is a genus of subfamily Hesperiinae, which has 18 members and is restricted geographically to the continental part of the southeastern Palaearctic and northern oriental regions. They have a thick body and relatively small wings. These wings are commonly dark brown or yellowish brown [1]. The main external features are characterized by the unspined mesotibia, the absence of a cell spot on the underside of each hind wing, and the serial, linear, and semi-hyaline spots. Male genitalia are distinguished by the elongated harpe, swollen tegumen, and bifid uncus [2].

2.1. The method of constructing and analyzing phylogenetic tree

The specimens from 15 of the estimated 18 species in the genus *Polytremis* were collected, from different localities. The DNA was isolated from leg tissue. The mitochondrial cytochrome c oxidase I (COI) gene, recommended as the universal and standard barcoding marker for species identification [3], was amplified approximately 490 bp. For nuclear DNA, three expansion segments of 18S rDNA and 28S rDNA were chosen, the slowly evolving genes used normally in higher classification studies [4]. The haplotype sequence matrix was used for all subsequent phylogenetic analyses. MEGA v4.0 was used to calculate the intra- and interspecific genetic divergences based on the K2P model [5]. Phylogenetic trees were constructed by the maximum-likelihood (ML) methods using PAUP 4.0b10 [6]. Relationships among the mitochondrial COI and concatenated sequence (mitochondrial COI + nuclear rDNA) haplotypes were represented as a haplotype network obtained by the software DnaSP4.90 [7] and Network4.5 using the median-joining method [8].

2.2. Genetic divergence, phylogenetics, and network of genus *Polytremis*

Figure 1A reveals five main clades and shows the ML tree based on the data set of COI. Clade I contained eight species: *Polytremis suprema* and *Polytremis gigantean, Polytremis caerulescens, Polytremis kiraizana* Sonan, 1938, and *Polytremis matsuii* Sugiyama, 1999 are first clustered with a strong support value, followed by clustering of *Polytremis gotama* and *Polytremis nascens* Leech, 1893. Then *Polytremis jigongi* Zhu, 2012 is revealed [2]. Two haplotypes of *P. jigongi* are clearly separated from other species form and strongly supported lineages. Clade II contains only *Polytremis theca*. It was reported to include three subspecies. They show a greater intraspecific genetic distance than some interspecific genetic distances in the genus *Polytremis*. Furthermore, the sister group relationship between *Polytremis zina* Evans, 1932 and *Polytremis pellucida* Murray, 1875 in Clade III is confirmed. Clade IV contains only one species, *Polytremis mencia* Moore, 1877. Clade V contains three species: *Polytremis discreta* Elwes & Edwards, 1897, *Polytremis lubricans* Herrich-Schäffer, 1869, and *Polytremis eltola* Hewitson, 1869, which are distributed sympatrically in the oriental region throughout Malaya and India.

http://dx.doi.org/10.5772/intechopen.69958

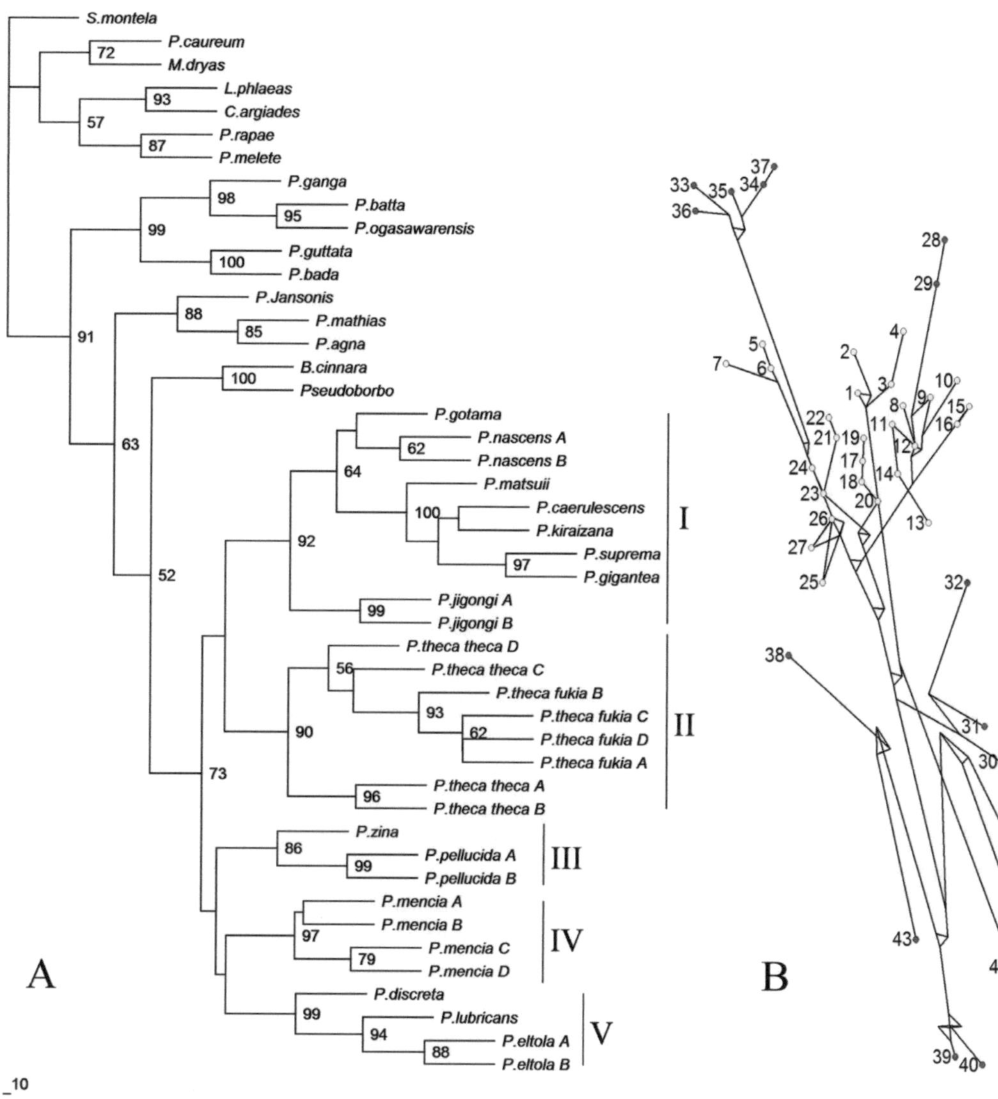

Figure 1. (A) Maximum-likelihood phylogeny on the basis of the mitochondrial COI sequences and (B) network on the basis of the mitochondrial COI sequences.

The level of DNA sequence divergence reflected the taxonomic hierarchy of the original species. The lowest intraspecific COI genetic distance was observed between *P. suprema* and *P. gigantea* (K2P distance 1.7%). Except for *P. theca*, the intraspecific distances were shorter. The COI data confirmed the sister group relationship between *P. suprema* and *P. gigantea*, which form a monophyletic group together with *P. matsuii*, *P. caerulescens*, and *P. kiraizana*. All of their interspecific distances were smaller than 3% (K2P distance). We can infer from morphological traits. These five species shared many morphological traits including ear-like uncus with a pair of processes, the absence of a cornuti, and thin coecum penis. *P. theca* was the only species for which the intraspecific genetic distance was greater than some interspecific genetic distances based on COI in the genus *Polytremis*. However, the distance was much less than the average interspecific genetic distances of the genus *Polytremis* (K2P distance 7.9%) [9]. *P. theca* was widely distributed in the south of the Qinling Mountains in China, except in the Hainan Province and the southern tropical regions of Yunnan Province [10]. Three subspecies of *P. theca* were reported on the basis of morphological features of the wings. Our specimens included two of them, namely, *Polytremis theca theca* and *Polytremis theca fukia*. The COI tree revealed these two distinct haplotype lineages without intermediates (K2P distance 4.2%). Additionally, the subspecies were separated by nuclear rDNA sequence, and the K2P distance was 0.3%, suggesting the possible existence of a sibling species paired with allopatric distribution.

The average interspecific rDNA genetic distance (K2P distance 1.0%) was far less than that of COI (K2P distance 7.9%). Except for *P. caerulescens* and *P. gotama,* other species in *Polytremis* could be distinguished with rDNA. These two species could be separated in the COI, and K2P distance was 1.9% but showed no variation in the rDNA. Based on mitochondrial and nuclear markers, the differences observed between results may contribute to recent separation, introgressive hybridization, or incomplete lineage sorting [11]. Because *Polytremis* is a fairly old genus and the splits of COI of the two lineages are also quite old, it seems that incomplete lineage sorting may not be an appealing explanation for the discordance. Additionally, *P. gotama* has been described as an independent species by COI data and morphological features [12]. A specimen of *P. gotama* and three specimens of *P. caerulescens* (from two populations) revealed two distinct haplotype lines without intermediates (K2P distances 4.9%). The observation indicates that they were two species based on the molecular and morphological level. Nevertheless, more specimens of the two species from different population need to be collected and analyzed in the future to see if this pattern is recovered consistently and further confirm the relationship of them. Instead, some arguments favor the assumption of recent separation [9]. As far as their morphological characteristics were concerned, *P. caerulescens* was considered to be closely related to *P. gotama* on the basis of the structural similarity of the cell spot on the upper side of the hind wing and the male genitalia, which were not found in the other *Polytremis* species [13]. Additionally, only these two species were observed and captured at altitudes higher than 2000 m. *P. gotama* was a little smaller than *P. caerulescens*. They both varied in other morphological traits which clearly support the existence of two closely related but distinct species, including male stigma on the upper side of the forewing and the ground color of the wings. rDNA showed no sequence variation, whereas K2P distances of the COI fragments reached 4.9%, suggesting a possible recent separation of *P. gotama* and *P. caerulescens*.

In *Lepidoptera,* thresholds have been proposed as 3% for COI [3]. The intraspecific and interspecific genetic divergences did not fall into separate intervals, and an obvious "barcode gap" did not occur in COI in our study of *Polytremis* [9]. It was entailed by two factors. Firstly, all intraspecific distances were less than 3% except for *P. theca*. However, we inferred *P. t. theca* and *P. t. fukia* could be a sibling species pair according the morphological and molecular data in the study. Secondly, the interspecific genetic distances among five sister species (*P. gigantean, P. suprema, P. caerulescens, P. kiraizana,* and *P. matsuii*) were less than 3%, which caused the interspecific and intraspecific genetic divergences to overlap from 1.7 to 3%. Regardless, for COI, the overlaps between intraspecific and interspecific variations would not affect identification in a thoroughly sampled evaluation [14]. The *Polytremis* species could be clustered with a well support and distinguished by tree-based methods, suggesting that the COI sequence could be used to correctly identify almost all species in the genus as a DNA barcode (**Figure 1A**). Additionally, the markers of the nuclear rDNA sequences used in our studies, three expansion segments of 18S and 28S rDNA, have been proposed as a reasonable alternative to mitochondrial COI. These markers could avoid problems of mitochondrial markers such as introgression and pseudogenes and identify or delimit species or species-like units as they are not inherited maternally [9].

2.3. Conclusion: combined morphological and molecular analysis

A total of 20 morphological characters yield a two-cluster solution with hierarchical cluster analysis. The first cluster includes 12 species of *Polytremis*, and the second includes the remaining 3 species and outgroups. All supported clades from the combined data matrix are also appearing in the molecular data matrix. ML tree based on COI constructed in this study showed that individuals belonging to the same species formed a monophyletic cluster. Meanwhile, there was considerable congruence in topology of the interspecies level for both mtDNA COI and concatenated sequences of ML trees indicating certain clades were well differentiated phylogenetically (**Figure 1**) [9]. The strong support for the monophyly of *Polytremis* was found in the analyses of the concatenated alignments and COI [9]. In genus *Polytremis*, the results obtained by hierarchical cluster analysis showed traditional classification was basically consistent with molecular phylogeny. However, because the morphological characters and character states were commonly homologous in *Polytremis*, the morphological analysis resulted in only limited resolution based on just 20 morphological characters. On contrary, molecular classification provided a lower artificial and more precise taxonomic rank [15]. Thus, the combination of the morphological and molecular matrix was better resolved for understanding of the phylogeny in the genus.

3. Taxonomic status of two sibling species of *Polytremis* (*Lepidoptera*: Hesperiidae)

The skipper *P. theca* is widely distributed in south China, except Taiwan, Hainan, and the southern tropical regions of Yunnan Province. Three subspecies have been recorded: *P. t. theca* [9] (west Sichuan and south Shaanxi Province), *P. t. fukia* [16] (Zhejiang to west Sichuan Province), and *Polytremis theca macrotheca* Huang, 2002 (Northwest Yunnan Province). In a preliminary study of molecular phylogeny of the genus *Polytremis* Mabille, 1904 using mitochondrial cytochrome c oxidase I (COI), we found the inter-subspecific distance between *P. t. theca* and *P. t. fukia* ranged up to 4.2%, which is higher than some interspecific genetic distances in *Polytremis*. Additionally, *P. theca* is also the only species whose intraspecific distance is more than 3%; thresholds of species identification have been proposed in *Lepidoptera* for COI, in genus *Polytremis* [9]. Thus, we suspected the possible existence of a sibling species paired or the cryptic diversity in the species.

3.1. Genetic divergences and haplotype networks

All 46 samples yielded high quality of DNA. A total of 19 haplotypes were identified in all 46 samples, and the haplotype network was constructed and presented in **Figure 2**. There was no shared haplotype among the four taxa. Haplotypes of the same taxon differed from each other by no more than five mutation distance. The five mutation distances existed between the haplotype Ptt I and Ptt III of *P. t. theca*. The potential ancestral haplotype of *P. t. fukia*, defined by its central position in the network, was designated as Ptf I, which was found in three samples from Tianmushan, one from Jinggangshan, and one from Anjiangping. Ptf II was the most common haplotype in *P. t. fukia* and shared with 10 samples. Haplotype Ptf III was found in two samples from Wuyishan. Haplotype Ptf XIII was identified in two samples from Maoershan and one sample from Anjiangping. The remaining haplotypes of *P. t. fukia* occurred in only one individual (**Figure 2A**).

The data set of nuclear *wingless* contains 390 nucleotide positions without gaps or stop codons, of which 18 positions are variable and 9 are parsimony informative. In total, 10 haplotypes were found in all samples, in which 2 haplotypes were found in *P. t. theca*, 4 in *P. t. fukia*, 3 in *P. nascens*, and 1 in *P. mencia*. Haplotypes of the same taxon differed from each other by no more than two mutation distances, in particular, the 2 haplotypes in 13 samples of *P. t. theca* differed by only one-mutation distance. Five nucleotide substitutions were observed between the potential ancestral haplotypes of *P. t. theca* (Ptt I) and that of *P. t. fukia* (Ptf I) (**Figure 2B**).

Overall, *P. t. theca* had a lower diversity than *P. t. fukia* according to the result of analysis of both mitochondrial COI and nuclear *wingless*. The haplotype diversity (Hd) and nucleotide diversity (π) for *P. t. theca* and *P. t. fukia* are given in **Table 1**. Additionally, they differed from each other by 5.07 ± 0.49% (4.3–5.9% divergence) for the COI sequences and by 1.70 ± 0.27% (1.3–2.1% divergence) for the *wingless* sequences.

3.2. Population structure and phylogenetic analysis

The analysis of molecular variance (AMOVA) for the COI sequences of *P. t. theca* and *P. t. fukia* revealed that 88.53% of the genetic variation was among populations and 11.47% was within populations (**Table 2**). The average ΦST value is 0.896 ($p < 0.01$), suggesting significant genetic variation among the populations. Pair-wise estimates of F_{ST} (0.885) and gene flow (Nm = 0.065) between *P. t. theca* and *P. t. fukia* suggest that the subspecies in this species are highly differentiated [16].

Mitochondrial haplotypes sampled from *P. theca* form well-supported clades that closely correspond with subspecific boundaries delimited primarily on the basis of wing color and pattern (**Figure 3A**). The haplotype clades associated with both subspecies are deeply genetically divergent, differing from each other by 5.07 ± 0.49% (4.3–5.9% divergence). This degree of divergence suggests that evolutionary separation of both subspecies occurred about 0.81 highest probablity density (HPD = 0.53–1.28) Mya, likely sometime during the Pleistocene based on a molecular clock calibration of 3.54% pair-wise divergence per million years for

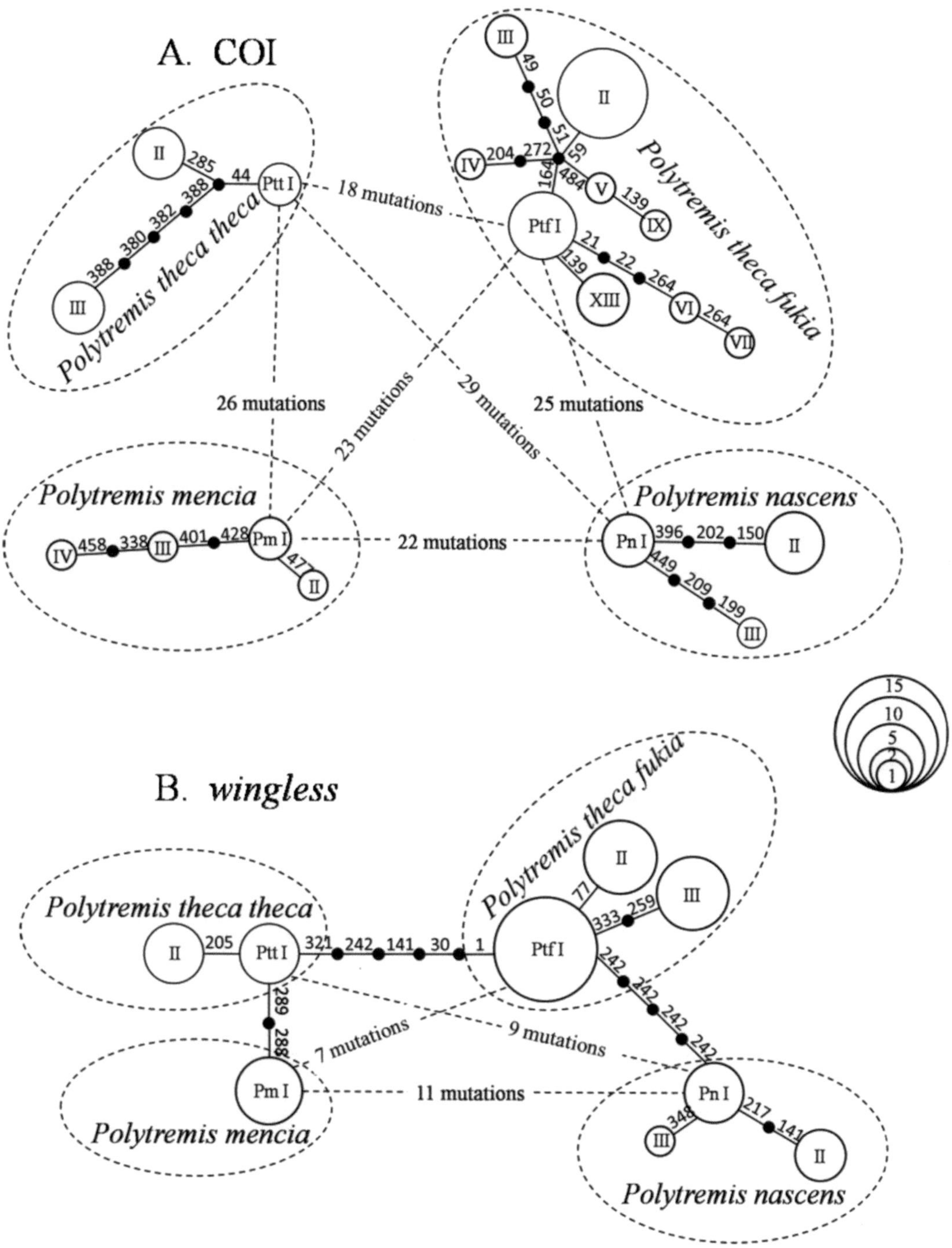

Figure 2. Network profile of (A) COI and (B) *wingless* gene haplotypes based on the nucleotide sequences of *P. t. theca*, *P. t. fukia*, *P. nascens*, and *P. mencia*.

	Ns	Nh	Hd	Nv	π	SD (π)	D	F
All *P. theca* samples (COI)	33	12	0.875	38	0.0207	0.0019	0.281	2.368
P. t. theca samples (COI)	8	3	0.750	6	0.0064	0.0048	1.598	2.631
P. t. fukia samples (COI)	25	9	0.803	13	0.0050	0.0070	−1.014	−1.886*
All *P. theca* samples (*wingless*)	33	5	0.773	9	0.0078	0.0057	1.122	1.509
P. t. theca samples (*wingless*)	8	2	0.571	1	0.0015	0.0010	1.444	1.100
P. t. fukia samples (*wingless*)	25	3	0.640	3	0.0030	0.0020	1.080	1.159

Ns, number of samples; Nh, number of haplotypes; Hd, haplotype diversity; Nv, number of variable sites; π, nucleotide diversity; SD, standard deviation; D, Tajima's D statistic; F, Fu's F statistic.
*Significant difference.

Table 1. Genetic diversity and neutrality tests calculated for *P. t. theca* and *P. t. fukia*.

a homologous mtDNA fragment in other insect species [17]. It is noteworthy for the nuclear *wingless* sequences that *P. t. theca* and *P. mencia* are considered distinct species with a genetic divergence of 0.65 ± 0.15%, while the *P. t. fukia* is currently considered a subspecies of the *P. t. theca* despite 1.70 ± 0.27% sequence divergence. The phylogenetic of *wingless* gene indicates that the *P. theca* is paraphyly with three species sisters to the clade of *P. t. theca* (**Figure 3B**). While 100% of *P. t. fukia* constitutes one separate clade, the clade consisting of *P. t. theca* also includes *P. pellucid, P. zina,* and *P. mencia*. The clade consisting of *P. t. theca* is not monophyletic, but complex. This suggests that *P. t. theca* and *P. t. fukia* differ from each other, as evident from the COI tree where they form two separate clades (**Figure 3A**). Concordance between strongly differentiated mtDNA, nuclear haplotype clades, and phenotypic variation supports the hypothesis that both subspecies of *P. theca* deserve recognition at the species level under the general lineage concept of species.

3.3. Demographic inference and estimation of divergence times

Demographic history changes were analyzed for *P. t. theca* and *P. t. fukia* populations through neutrality tests and mismatch distribution. The neutrality tests reveal that the mitochondrial COI appear to be not evolving neutrally as Fu's F values in *P. t. fukia* group are negative significantly (**Table 1**). The Tajima's D and Fu's F values were nonsignificantly positive in

Source of variation	*df*	Sum of squares	Variance components	Percentage variation	Φ statistic
Among populations	1	169.650	9.98266 Va	88.53	–
Within populations	11	45.269	1.29341 Vb	11.47	0.896 ($p < 0.01$)
Total	12	214.919	11.27607		
Fixation index	0.8853				

Table 2. Analysis of molecular variance (AMOVA) for the COI sequences of *P. t. theca* and *P. t. fukia*.

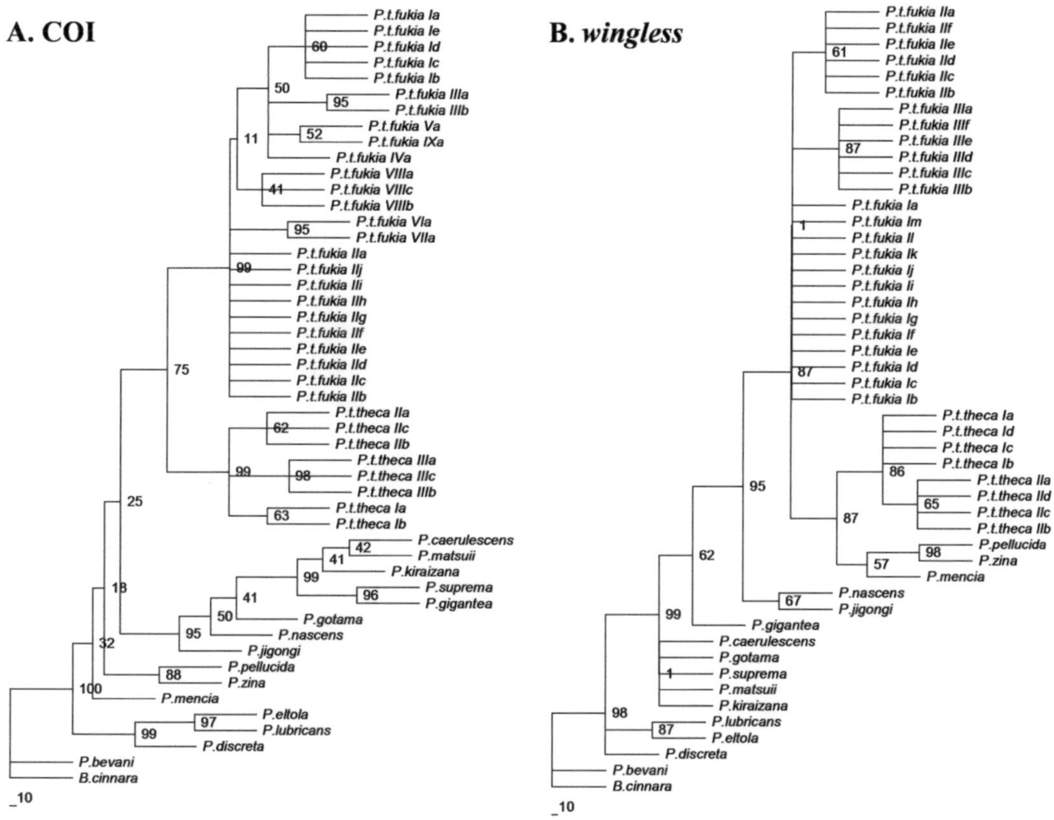

Figure 3. The maximum-likelihood tree for (A) mitochondrial COI and (B) nuclear *wingless* haplotypes of *Polytremis*.

P. t. theca group and all *P. theca* sample group. The mismatch analysis yielded a unimodal distribution of pair-wise differences for *P. t. fukia* compared to the multimodal distribution of *P. t. theca* samples and the pooled samples. According to Rogers and Harpending [18], the observed curves with unimodal represent population expansion and the observed curves with many peaks or resemblance to expected curves mean equilibrium population, which further elucidates the demographic history of *P. theca*. The results suggest population expansion in *P. t. fukia* and population equilibrium in *P. t. theca*. We still confirm the result of the population size change in haplotype network (**Figure 2**). Statistical parsimony network reflects genealogical relationships of the mtDNA haplotypes, that is, single mutation steps separate adjacent haplotypes in the network and older haplotypes are placed at internal branching points, whereas younger ones occur toward the tip positions [19]. The haplotype network of *P. t. fukia* displays a star-like pattern (**Figure 2**). Haplotype I, the second most common and geographically widespread in central-west of China, is in the star's center, and derivatives are connected to it by short branches [16]. Based on coalescence theory, the star-like topologies for this cluster strongly suggest the effect of a population expansion [20]. Divergence time analysis with an uncorrelated lognormal relaxed clock run in Bayesian MCMC analysis of molecular sequences (BEAST) produced a tree with a topology similar to ML tree (**Figure 4**).

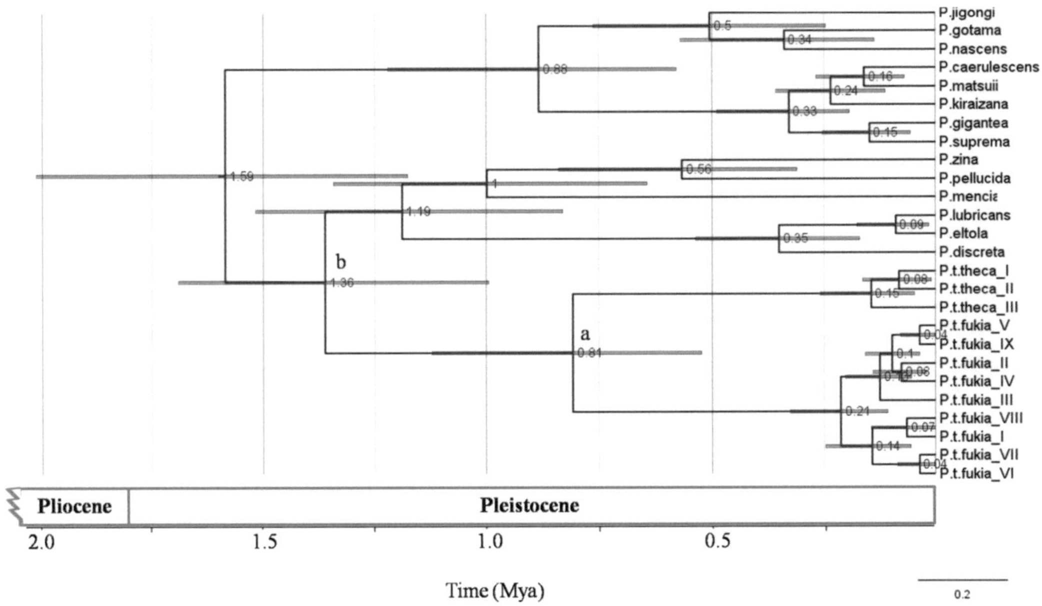

Figure 4. Bayesian inference (BI) tree of mtDNA datasets for *Polytremis* using uncorrelated lognormal relaxed clock.

P. t. theca diverged from *P. t. fukia* around 0.81 (HPD = 0.53–1.28) million years ago (Mya) during the Pleistocene (node a in **Figure 4**). *P. theca* diverged from other congeners included in the analysis about 1.36 (HPD = 1.02–1.53) Mya during the Pleistocene eras (node b in **Figure 4**). In our study, a higher F_{ST} value indicated a lower level of gene flow (Nm) and higher genetic differentiation among populations. The results of two-level AMOVA show that significant genetic variation exists among the examined populations. These results provide a second line of support to a conclusion that the *P. t. fukia* is a different species [16].

3.4. Conclusion

There is a small region of overlap in west Sichuan province in the distribution of *P. t. theca* and *P. t. fukia*, but otherwise they are not sympatric. *P. t. theca* inhabits the higher elevations of west Sichuan and south Shaanxi Province [21]. *P. t. fukia* occurs in the whole Southern China except Taiwan, Hainan, and south Yunnan Province, from Zhejiang to west Sichuan Province [22]. According to the description on morphological variation between *P. t. theca* and *P. t. fukia* [23], we found a different morphological feature existing in female genitalia except for the different colors and spot numbers in some part of wings (**Table 3**). The lateral edge of lamella antevaginalis of female *P. t. fukia* is more rounded than that of *P. t. theca*. We suspected the taxonomic status of the subspecies from their geographic separation and the morphological variation. Our investigations and analyses revealed significant molecular and biogeographical differences between *P. t. theca* and *P. t. fukia*. We propose that *P. t. fukia* should be treated as a distinct species called *Polytremis fukia* [16] under the phylogenetic species concept. In fact, it has been

	Wing					Genital
	Color of cilia of wings	**Color of underside ground**	**Number of spots in space Cu2 of the forewing**	**Color of scales scattered in costa and subapical area of forewing**	**Color of scales scattered in discal area and dorsum of hind wing**	**Ductus bursae**
P. t. theca	Brown	Yellowish brown	0	Greenish ochreous	Greenish ochreous	Thin
P. t. fukia	Grayish white	Greenish ochreous	1 or 2	Grayish white	Grayish white	Thick

Table 3. Different morphological features of genitals and wings between *P. t. theca* and *P. t. fukia*.

recently found that other species previously considered subspecies based on morphology are in fact sibling species that passed unnoticed until the advent of molecular techniques [24]. Results from our study strengthen information about the *Polytremis* species complex and help in developing appropriate integrated pest management strategies for these insect pests.

4. The application of mitochondrial genome and microsatellite DNA in *Polytremis*

With the development of the research, the single molecular fragment cannot meet the research requirements. New molecular markers need to be explored. Recently, the mitochondrial genome has become one of the important molecular markers to explore different categories of *Lepidoptera*. Additionally, the ideal molecular marker, microsatellite DNA, has become very prevalent in molecular ecological studies including inferring population genetic structure, exploring taxonomic status, and studying reproductive ecology.

4.1. Complete mitochondrial DNA genome of *P. nascens* and *Polytremis jigongji*

Lepidoptera is the second largest insect order in the world and contains more than 160,000 species. However, the information currently existing on lepidopteran mitogenomes is limited. To date, only 236 complete or nearly complete mitogenome sequences have been determined to belong to six superfamilies. The phylogenetic inference based on the variations at such short gene sequences is not always robust and may lack sufficient resolution compared to the phylogenies based on longer mitogenome sequences [25]. Additionally, mitogenomes are also applied to the studies on comparative and evolutionary genomics [26], molecular evolution [27], phylogeography [28], etc. Thus, further insight into *Lepidoptera*n phylogeny and evolution awaits more related species sequences to be determined [29].

The complete mtDNA genome of *P. nascens* was a circular molecule of 15,392 bp in length, including the standard 13 protein-coding genes (PCGs), 2 ribosomal RNA (rRNA) genes, 22 transfer RNA (tRNA) genes, and 1 noncoding region. The overall-based composition was 39.7% A, 40.7% T, 7.7% G, and 11.9% C, with a sight A+T bias of 80.4%. Thirteen PCGs and two rRNA were first identified using an open reading frame (ORF) finder, specifying the invertebrate mitochondrial genetic code. Then, they were calibrated by sequence similarity using published lepidopteran mitogenome sequences. All PCGs use standard ATN (ATT, ATG, or ATA) as the start codon except COX1 that uses CGA. Eight PCGs (ND2, ATP8, ATP6, COIII, ND3, ND4, ND6, and ND1) employ the typical stop codon TAA, while the remaining five PCGs terminate with a single T 33 [22].

The complete mtDNA genome of *P. jigongi* was a circular molecule of 15,353 bp in length, including the standard 13 protein-coding genes (PCGs), 2 ribosomal RNA (rRNA) genes, 22 transfer RNA (tRNA) genes, and 1 noncoding region. Its organization and arrangement are identical to other skippers [30, 31]. The overall-based composition was 39.8% A, 41.1% T, 7.6% G, and 11.5% C, with a sight A+T bias of 80.9%. Thirteen PCGs and two rRNA were identified using an open reading frame (ORF) finder and calibrated by sequence similarity using published lepidopteran mitogenome sequences. All PCGs use standard ATN (ATT, ATG, or ATA) as the start codon except COX1 that uses CGA. Eight PCGs (ND2, ATP8, ATP6, COIII, ND3, ND4, ND6, and ND1) employ the typical stop codon TAA, while the remaining five PCGs terminate with TA or T [22]. Alignment of amino acid sequences of each of individual 13 PCGs was performed through Clustal X [32], and the phylogenetic analysis was carried out using neighbor-joining (NJ) method with MEGA version 5.0 program [33] (**Figure 5**).

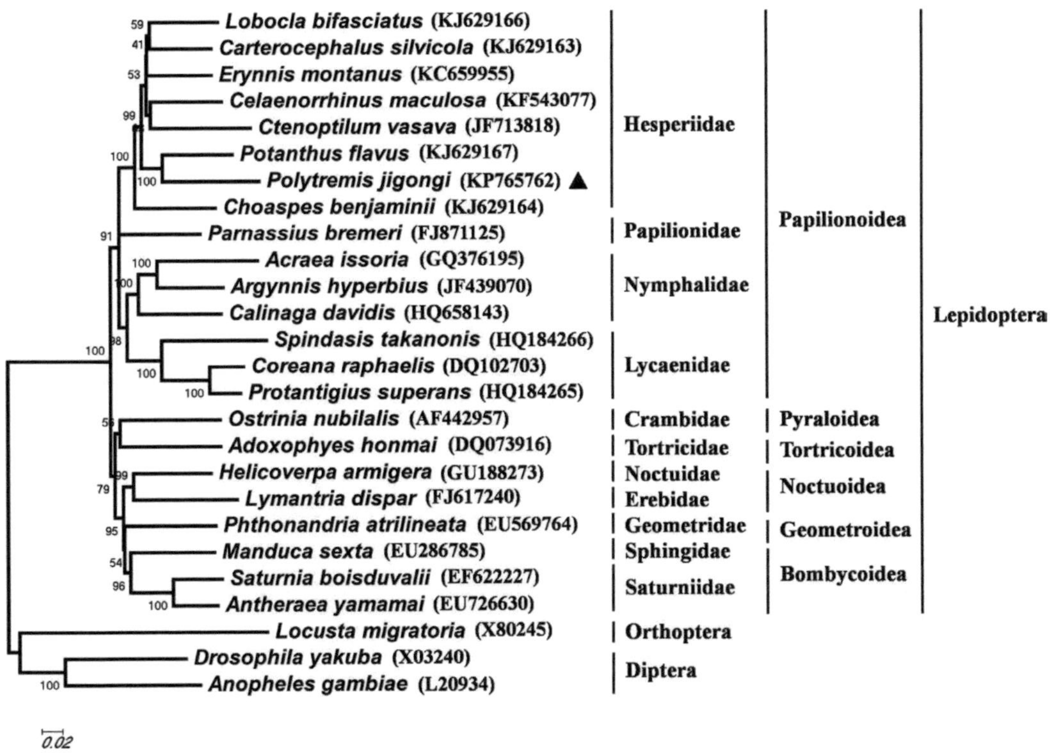

Figure 5. Phylogenetic tree of the lepidopterans based on 13 PCG nucleotide sequences of the mitogenome using NJ analysis.

4.2. Isolation and characterization of microsatellite loci in *P. nascens* and *P. fukia*

Microsatellites are highly polymorphic and codominant molecular markers based on simple repeated and frequent sequences common in the all-living organisms, which have proven to be a powerful tool available in population genetic and evolutionary studies [34]. The ideal molecular marker has become very prevalent in studies of insects over the last 10 years [35]. Variability of the 12 polymorphic microsatellites was surveyed in 53 individuals of *P. nascens*. We found 53 different multilocus genotypes, and the number of alleles per locus ranged from 3 to 12. Observed (HO) and expected heterozygosity (HE) values ranged from 0.33 to 0.71 and from 0.61 to 0.90. Only one locus (Polynyas 13) ($p < 0.01$) that deviated from HWE showed significant heterozygote deficit in the populations of Lianglu, Shengtangshan, and Maoershan [33]. Analysis performed with Micro-Checker showed that the deviation was attributed to the homozygote excess with null allele frequency of 0.2041, 0.2914, and 0.3533 in the three populations, respectively [33]. No linkage disequilibrium was detected for any pair of loci ($p > 0.01$) in any populations following Bonferroni correction [33]. Comparisons of pair-wise F_{ST} among six populations show the genetic difference based on 12 microsatellite loci (**Table 4**). The population of Baishanzu showed the largest pair-wise F_{ST} values, corresponding to relatively geographic isolation from the other five populations [33].

The variability of the 11 polymorphic microsatellites was surveyed in 21 individuals of *P. fukia*. We found 21 different multilocus genotypes, and the number of alleles per locus ranged from 5 to 10. Observed heterozygosity (*Ho*) values ranged from 0.48 to 0.65 and expected heterozygosity (*He*) values ranged from 0.69 to 0.86. Polymorphic information content (PIC) ranged from 0.59 to 0.88 per locus, and all markers were highly informative (PIC > 0.5) [33]. All loci were in Hardy-Weinberg equilibrium, consistent with inbreeding and/or Wahlund effects. No linkage disequilibrium was detected in any pair of loci [33]. Micro-Checker software found no evidence of scoring error due to stuttering or large allele dropout. We also tested the selected primers for amplification on eight other *Polytremis* species: *P. caerulescens* Mabille, *P. jigongi* Zhu, *P. theca* Evans, *P. zina* Evans, *P. mencia* Moore, *P. lubricans* Herrich-Schäffer, *P. eltola* Hewitson, and *P. discrete* Elwes & Edwards. Of the 11 markers, we found that cross-amplification was successful for 10 loci in at least one conge-

	Baishanzu	**Hailuogou**	**Lianglu**	**Shengtangshan**	**Maoershan**	**Kuankuoshui**
Baishanzu						
Hailuogou	0.044					
Lianglu	0.050	0.003				
Shengtangshan	0.029	0.025	0.022			
Maoershan	0.023	0.021	0.020	0.013		
Kuankuoshui	0.026	0.012	0.014	0.022	0.017	

Table 4. Comparisons of multilocus pair-wise F_{ST} values ($p < 0.05$) among 6 regional populations of *P. nascens* based on 12 microsatellite loci.

neric species [33]. Two primer loci amplified satisfactorily among all congeneric species, indicating that these loci may be useful for population genetics, including phylogeography, species cohesion and delimitation, and barriers to gene flow for other congeneric and related species [33].

5. *Wolbachia* infection and influence in *Polytremis* species

Wolbachia may be the most widespread endosymbiont in terrestrial ecosystems. The maternally inherited endosymbiotic bacteria infect perhaps two-thirds of present-day insect species [36, 37]. The transmission of *Wolbachia* is primarily vertical and secondarily horizontal [38]. The bacteria manipulate the reproduction of their host to ensure their vertical transmission by cytoplasmic incompatibility, feminization, male killing, and parthenogenesis [39]. *Wolbachia* can potentially influence mitochondrial variation of their hosts. The linkage disequilibrium is expected to occur between them since they are co-transmitted maternally. The rapidly spread of *Wolbachia* in the host populations can result in the hitch-hiking effect of mitochondrial DNA [31]. One particular mitochondrial haplotype can reduce mtDNA polymorphism in the infected population and sweep through a population [19]. *Wolbachia* may also drive introgression of mtDNA following hybridization events between sibling species: the introduction and spread of *Wolbachia* in a novel species result in spread of the mtDNA from the neighboring species [40].

Wolbachia infections have been reported in various *Lepidoptera* families such as Papilionidae, Lycaenidae, Pieridae, Nymphalidae, Hesperiidae, Pyralidae, Noctuidae, and Lasiocampidae. A few butterfly species harboring the bacteria have been thoroughly studied. We have reported the molecular phylogeny of the genus in a prior study [9]. Meanwhile, we have preliminarily screened for the presence of *Wolbachia* in *Polytremis* and found at least three species (*P. nascens* Leech, 1893, *P. theca* Evans, 1937 and *P. pellucid* Murray, 1875) are infected with *Wolbachia*.

5.1. *Wolbachia* infection status and genetic structure in natural populations of *P. nascens*

We surveyed *Wolbachia* infection rates in 14 regional populations of *P. nascens*. Twenty-one specimens (31%) infected with *Wolbachia* in seven regional populations were detected. However, all specimens are free from infection with *Wolbachia* in the other seven regional populations. The infection rates between the female and male butterflies show no significant difference ($\chi^2 = 0.65$, $p > 0.05$). The further rearing experiment would be made to reveal whether there was a sex-ratio distortion in *P. nascens* induced by the strain of *Wolbachia*. A positive relationship of *Wolbachia* infections and latitudinal distribution has also been found. The lower or no *Wolbachia* infection rates in central-west of China may indicate that incidence is apparently lower in regions experiencing longer dry seasons and higher average daily temperatures. The higher *Wolbachia* prevalence occurs in more southerly moist and temperate populations (**Figure 6**). This has been observed in the beetle *Chelymorpha alternans* and in ants of the genus *Solenopsis* [41, 42].

http://dx.doi.org/10.5772/intechopen.69958

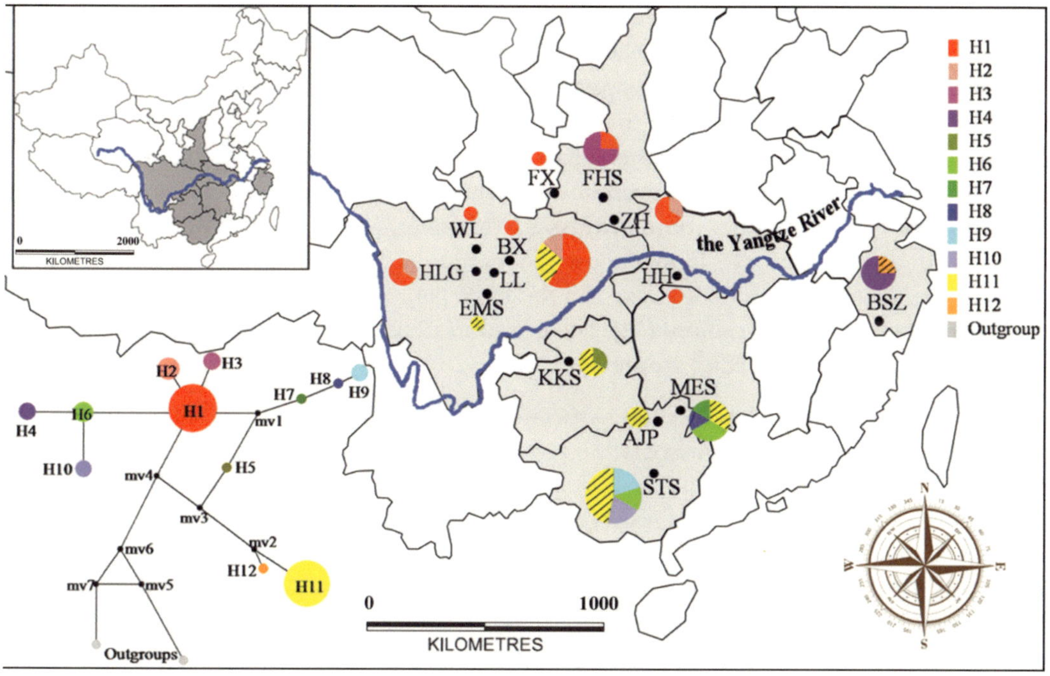

Figure 6. Distribution of mtDNA haplotypes among populations of *P. nascens* collected in China.

Many explanations have been proposed that deviation from neutral evolution in a species and the absence of diversity in mtDNA can be associated with either a genome-wide bottleneck effect or a selective sweep on mtDNA [43]. In our study, the uninfected butterflies show higher mtDNA polymorphism than butterflies infected with *Wolbachia* (**Table 5**). The perfect concordance of *Wolbachia* infection status and mtDNA polymorphism suggests that the mitochondrial genetic structure of the host insects may be strongly affected by the *Wolbachia* infection. Decreased mtDNA polymorphism as a consequence of *Wolbachia* infection has also been reported in several other insects [38, 40]. We also found that the mtDNA genes of the butterflies infected with *Wolbachia* deviated significantly from neutral evolution according to both D (−2.3303, $p < 0.05$) and F values (−3.7068, $p < 0.05$), while this was not so for uninfected ones [31]. These results suggest that the populations of *P. nascens* have

	N	Number of haplotypes	Haplotype diversity	Number of variable sites (S)	π	SD (π)	D	F
All sequences	67	12	0.797	119	0.0197	0.0012	1.4795	2.0907*
Infection	21	2	0.095	12	0.0006	0.0005	−2.3303*	−3.7068*
Free from infection	46	10	0.750	77	0.0122	0.0015	0.8835	1.9320

N, number of samples; π, nucleotide diversity; SD, standard deviation; D, Tajima's D statistic; F, Fu's F statistic.
*Significant difference.

Table 5. Mt haplotype and nucleotide diversity estimates from infected and uninfected samples.

recently been subjected to a *Wolbachia*-induced sweep, making the mtDNA undergo purifying selection. Additionally, we analyzed the population size change of *P. nascens* infected with *Wolbachia* and all *P. nascens*, respectively, by the software DnaSP4.90 [7] and got no evidence for population expansion in *Wolbachia*-infected group [18]. However, we got unimodal curves representing population expansion in *P. nascens*. A selective sweep would erase variability in the population even under population expansion, potentially eliminating any evidence of past demographic processes. Moreover, recent studies also revealed that there are some other endosymbionts known to manipulate host reproduction like *Wolbachia*, for example, *Arsenophonus*, *Cardinium*, and *Rickettsia* [44]. A wide range of insect species can host more than one endosymbiont [42]. The infection status of the secondary endosymbionts in *P. nascens* needs to be further studied.

Although there seems to be a strong association existed between mtDNA haplotypes and *Wolbachia* infection status, the association between mtDNA haplotypes, *Wolbachia* infection, and geographical distribution is weak [31]. In our preliminary experiment, we found that two sympatrically distributed sister species of *P. nascens* (*P. theca* and *P. pellucid*) are infected with *Wolbachia*. We could not eliminate the possibility of the multiple introgression events and hybridizations between the species pairs [40]. Thus, we excluded the infected group (Clade I) in the analysis of biogeographical implications of *P. nascens*. It is notable that if we only take the uninfected group into account, the conclusion can be drawn that *P. nascens* probably has two genetically diverse and geographically localized clades in China based on the mtDNA haplotype phylogeny and networks (**Figure 6**) [31]. They are the central-west of China clade (H1–H3) and the Eastern and Southern China clade (H4–H10). These two clades are isolated mainly by the Yangtze River, with the exception of a specimen in HH [31]. Our results were similar to those obtained from the striped stem borer, *Chilo suppressalis* [45], and the melon fly, *Bactrocera cucurbitae* [46], which suggest that the Yangtze River Range has acted as a substantial barrier to gene flow. This contrasts with studies of the beet armyworm, *Spodoptera exigua*, and the migratory locust, *Locusta migratoria*, and the cotton bollworm *Helicoverpa armigera*, which showed little or no evidence that the Yangtze River Range limits gene flow [47–49].

The network analysis reflects genealogical relationships of the mtDNA haplotypes [31]. The single mutation steps separate adjacent haplotypes in the network and older haplotypes are placed at internal branching points, whereas younger ones occur toward the tip positions [50]. The haplotype network of *P. nascens* displays a star-like pattern (**Figure 6**). Haplotype 1 (H1) is in the star's center, and derivatives are connected to it by short branches. The haplotype is the most common and geographically widespread in central-west of China. The star-like topologies for this cluster strongly suggest the effect of a population expansion based on coalescence theory [20]. We still confirm the result of the population size change of *P. nascens* with the software DnaSP4.90 and got unimodal curves representing population expansion [31]. The most common H1 had strong support as the ancestral haplotype due to its representation in a significant proportion of individuals in all populations and its basal location in the network [31]. A reticulation connecting multiple haplotypes from the Eastern and Southern China clade was formed in the network [31].

5.2. A prevalence survey of *Wolbachia* in *P. fukia*

Of the butterflies examined by diagnostic PCR for 16S rRNA, 47% (15/32) were *Wolbachia* positive. The infection rates in female and male are 69% (11/16) and 25% (4/16). For characterization of *Wolbachia* strains, we amplified a segment of the *Wolbachia* cell cycle gene *ftsZ*. BLAST searches of the *ftsZ* sequences yielded a significant sequence similarity between the *Wolbachia* strains infecting *P. fukia* and the *Wolbachia* supergroup A typically found in insects. Phylogenetic analyses suggested that the bacteria are subdivided into two strains with a genetic divergence of approximately 2.8%. These strains will be referred to as *w*Fuk1 and *w*Fuk2 in the following.

Figure 7A shows the ML tree based on the data set of the concatenated sequences and supports the monophyly of *P. fukia*. On the phylogeny, the sequences are split into three clades supported by high bootstrap values. Clade I exclusively consists of the *P. fukia* from two geographical populations in Southwest side of their distribution area (Maoershan and Anjiangpin). These butterflies are either free from *Wolbachia* infection or infected with *w*Fuk2. Clade II exclusively consists of the butterflies from four geographical populations (West Tianmushan, Qingliangfeng, Wuyishan, and Jinggangshan), which are either free from *Wolbachia* infection or infected with *w*Fuk2. These two clades are consistent with the distribution of geographical population. Eight females constitute Clade III are invariably infected with *w*Fuk1. In our data set comprising 32 individuals (16♀ and 16♂), we did not detect any male *P. fukia* infected

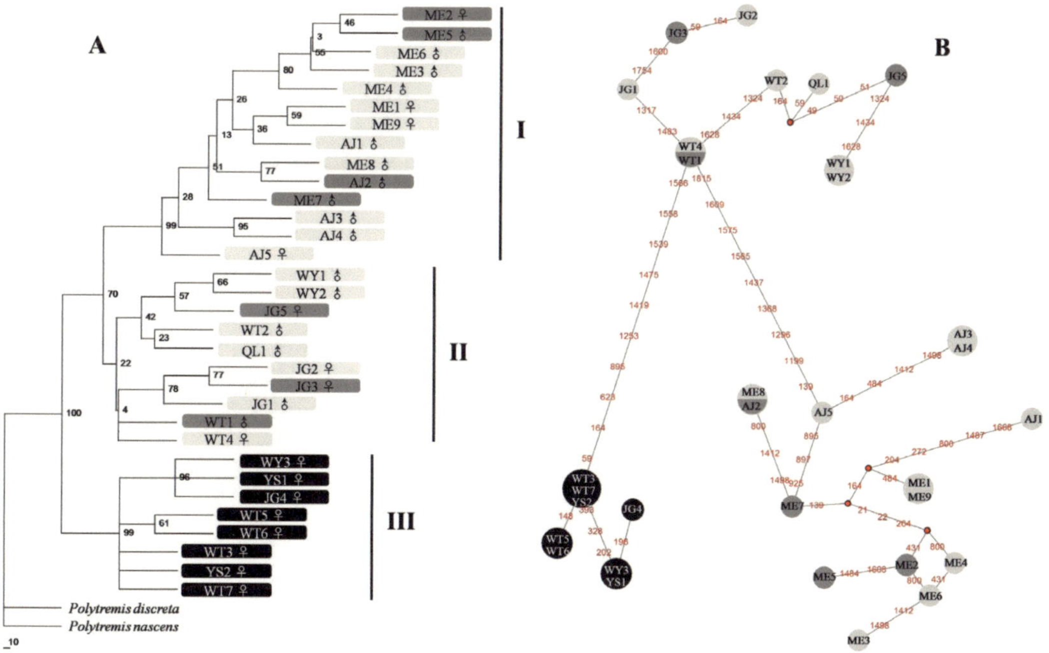

Figure 7. (A) Maximum-likelihood phylogeny on the basis of the concatenated mitochondrial sequences and (B) network on the basis of the concatenated mitochondrial sequences constructed with software Network4.5.

	N	Number of haplotypes	Hd	SD (Hd)	Number of variable sites (S)	π	SD (π)	D	F
All sequences	32	23	0.980	0.012	50	0.008	0.001	0.568	−3.913
Infection with *w*Ful1	8	4	0.821	0.101	5	0.001	0.000	0.840	0.428
Infection with *w*Ful2	7	7	1.000	0.076	32	0.008	0.001	0.077	−1.085
Free from infection	17	14	0.978	0.027	36	0.007	0.002	0.522	−2.510

N, number of samples; Hd, haplotype diversity; π, nucleotide diversity; SD, standard deviation; D, Tajima's D statistic; F, Fu's F statistic.

Table 6. Mt haplotype and nucleotide diversity estimates from infected and uninfected samples.

with *w*Fuk1. All eight individuals infected with *w*Fuk1 are female. Males belong to Clade I or Clade II and, if infected, the strains *w*Fuk2. All mtDNA of *P. fukia* were used for network construction with the software Network4.5 using the median-joining method (**Figure 7B**). Twenty-three haplotypes are clustered into three groups in accordance with three clades in ML tree (see **Figure 7A**). Two haplotypes were shared by the individuals infected with *Wolbachia* and free from *Wolbachia*. The *Wolbachia*-induced sweep has been shown in *P. nascens* [51]. However, such *Wolbachia* effects on mtDNA variation in *P. fukia* examined in this study were not as conspicuous as in the above studies. Still, mtDNA variation in the *P. fukia* is likely to be weakly associated with the presence of *w*Fuk1, although the sample size was not large enough to draw a firm conclusion [37]. The nucleotide diversity in *w*Fuk1-infected butterflies is smaller than that of uninfected butterflies (**Table 6**) and their mt concatenated sequences form a clade solely (**Figure 7**), presumably reflecting a *Wolbachia* sweep. This tendency was consistently seen in all the mitochondrial regions examined. A larger sample will reveal more in detail the association of *Wolbachia* infection status with variation of mtDNA in *P. fukia*.

Acknowledgements

This study was financially supported by grants from National Natural Science Foundation of China (no. 31401997) and Young Teacher Training Scheme for Colleges and Universities in Shanghai (no. ZZssd15068).

Author details

Weibin Jiang

Address all correspondence to: jiangwb@shnu.edu.cn

College of Life and Environmental Sciences, Shanghai Normal University, Shanghai, People's Republic of China

References

[1] Huang H. Some new butterflies from China-2. Atalanta. 2002;**33**:109-122

[2] Zhu JQ, Chen ZB, Li LZ. *Polytremis jigongi*: A new skipper from China (*Lepidoptera*: *Hesperiidae*). Zootaxa. 2012;**3274**:63-68

[3] Hebert PDN, Penton E, Burns J. Ten species in one: DNA barcoding reveals cryptic species in the neotropical skipper butterfly *Astraptes fulgerator*. Proceedings of National Academy Sciences. 2004;**101**:14812-14817

[4] Hwang UW, Ree HI, Won K. Evolution of hypervariable regions, V4 and V7, of insect 18S rRNA and their phylogenetic implications. Zoological Science. 2000;**17**:111-121

[5] Tamura K, Dudley J, Nei M, Kumar S. MEGA4: Molecular evolutionary genetics analysis (MEGA) software version 4.0. Molecular Biology and Evolution. 2007;**24**:1596-1599

[6] Swofford DL. PAUP* Phylogenetic Analysis Using Parsimony (*and other methods). Version 4.0b10 ed. Massachusetts: Sunderland Sinauer Associates; 2002

[7] Rozas J, Sánchez-Delbarrio JC, Messeguer X, Rozas R. DnaSP, DNA polymorphism analyses by the coalescent and other methods. Bioinformatics. 2003;**19**:2496-2497

[8] Bandelt HJ, Foster P, Röhl A. Median-joining networks for inferring intraspecific phylogenies. Molecular Biology and Evolution. 1999;**16**:37-48

[9] Jiang WB, Zhu JQ, Song C, Li XY, Yang Y, Yu WD. Molecular phylogeny of the butterfly genus *Polytremis* (*Hesperiidae*, *Hesperiinae*, Baorini) in China. PLoS ONE. 2013;**8**:e84098

[10] Huang H. A list of butterflies collected from Nujiang and Dulongjiang, China with descriptions of new species, new subspecies and revisional notes. Neue Entomologische Nachrichte. 2003;**55**:3-114

[11] Raupach MJ, Astrin JJ, Hannig K, Peters MK, Stoeckle MY, Wägele JW. Molecular species identification of Central European ground beetles (*Coleoptera*: *Carabidae*) using nuclear rDNA expansion segments and DNA barcodes. Frontiers in Zoology. 2010;**7**:26

[12] Sugiyama H. New butterflies from western China. Pallarge. 1999;**6**:1-14

[13] Zhu JQ. A Taxonomic Study on the Tribe Baorini Doherty, 1886 (*Lepidoptera*: *Hesperiidae*: *Hesperiinae*) of China. Shanghai: Shanghai Normal University. Dissertation; 2012

[14] Meyer CP, Paulay G. DNA barcoding: Error rates based on comprehensive sampling. PLoS Biology. 2005;**3**:2229-2238

[15] Tautz D, Arctander P, Minelli A, Thomas RH, Vogler AP. A plea for DNA taxonomy. Trends in Ecology and Evolution. 2003;**18**:70-74

[16] Jiang WB, He HY, Li YD, Ren MY, Ma YZ, Zheng LL, Zhu JQ, Yu WD*. Taxonomic status and molecular phylogeography of two sibling species of *Polytremis* (*Lepidoptera*: *Hesperiidae*). Scientific Reports. 2016;**6**:20820

[17] Papadopoulou A, Anastasiou I, Vogler AP. Revisiting the insect mitochondrial molecular clock: The mid-Aegean trench calibration. Molecular Biology and Evolution. 2010;**27**:1659-1672

[18] Rogers AR, Harpending H. Population growth makes waves in the distribution of pairwise genetic differences. Molecular Biology and Evolution. 1992;**9**:552-569

[19] Narita S, Nomura M, Kato Y, Fukatsu T. Genetic structure of sibling butterfly species affected by *Wolbachia* infection sweep: Evolutionary and biogeographical implications. Molecular Ecology. 2006;**15**:1095-1108

[20] Slatkin M, Hudson RR. Pairwise comparisons of mitochondrial DNA sequences in stable and exponentially growing populations. Genetics. 1991;**129**:555-562

[21] Chou I. In: Monographia Rhopalocerorum Sinensium. Zhengzhou: Henan Scientific and Technological Press; 1994

[22] Jiang WB, Zhu JQ*, Yang QC, Zhao HD, Chen MH, He HY, Yu WD*. Complete mitochondrial DNA genome of *Polytremis nascens* (*Lepidoptera*: *Hesperiidae*). Mitochondrial DNA Part A. 2016;**27**(5):3131-3132

[23] Eliot JN, Corbet AS, Pendlebury HM. In: The butterflies of the Malay Peninsula. 4th ed. Kuala Lumpur: United Selangor Press; 1992

[24] Evans III EL, Martinez-Gomez JE, Sehgal RNM. Phylogenetic relationships and taxonomic status of the endemic Socorro Warbler (*Setophaga pitiayumi graysoni*). Journal of Ornithology. 2015;**156**:363-370

[25] Powell AFLA, Barker FK, Lanyon SM. Empirical evaluation of partitioning schemes for phylogenetic analyses of mitogenomic data: an avian case study. Molecular Phylogenetics and Evolution. 2013;**66**:69-79

[26] Ballard JWO. Comparative genomics of mitochondrial DNA in *Drosophila simulans*. Journal of Molecular Evolution. 2000;**51**:64-75

[27] Zakharov EV, Caterino MS, Sperling FAH. Molecular phylogeny, historical biogeography, and divergence time estimates for swallowtail butterflies of the genus *Papilio* (*Lepidoptera*: *Papilionidae*). Systematic Biology. 2004;**53**:193-215

[28] Malyarchuk B, Derenko M, Denisova G. A mitogenomic phylogeny and genetic history of sable (*Martes zibellina*). Gene. 2014;**550**:56-67

[29] Wang JP, Cao TW, Xuan SB, Wang H, Zhang M, Ma EB. The complete mitochondrial genome of *Sasakia funebris* (Leech) (*Lepidoptera*: *Nymphalidae*) and comparison with other Apaturinae insects. Gene. 2013;**526**:336-343

[30] Hao J, Sun Q, Zhao H, Sun X, Gai Y, Yang Q. The complete mitochondrial genome of *Ctenoptilum vasava* (*Lepidoptera*: *Hesperiidae*: *Pyrginae*) and its phylogenetic implication. Comparative and Functional Genomics. 2012;**2012**:328049. DOI: 10.1155/2012/328049

[31] Jiang WB, Zhu JQ, Chen MH, Yang QC, Du X, Chen SY, Zhang LN, Yu YM, Yu WD. *Wolbachia* infection status and genetic structure in natural populations of *Polytremis nascens* (*Lepidoptera*: *Hesperiidae*). Infection, Genetics and Evolution. 2014;**27**:202-211

[32] Thompson JD, Gibson TJ, Plewniak F, Jeanmougin F, Higgins DG. The CLUSTAL_X windows interface: Flexible strategies for multiple sequence alignment aided by quality analysis tools. Nucleic Acids Research. 1997;**25**:4876-4882

[33] Jiang WB, Chen MH, He HY, Wu YJ, Zhu JQ*, Yu WD*. Eleven microsatellite loci developed from the Chinese Hesperid butterflies, *Polytremis fukia* (*Lepidoptera*: *Hesperiidae*), and cross-amplification in other species of *Polytremis*. Applied Entomology and Zoology. 2016;**51**(4):669-672

[34] Goldstein D, Schlötterer C. Microsatellites: Evolution and Applications. Oxford: Oxford University Press; 1999

[35] Zakharov EV, Lobo NF, Nowak C, Hellmann JJ. Introgression as a likely cause of mtDNA paraphyly in two allopatric skippers (*Lepidoptera*: *Hesperiidae*). Heredity. 2009;**102**:590-599

[36] Hilgenboecker K, Hammerstein P, Schlattmann P, Telschow A, Werren JH. How many species are infected with *Wolbachia*?—A statistical analysis of current data. FEMS Microbiology Letters. 2008;**281**:215-220

[37] Jiang WB, Wu YJ, He HY, Zhu JQ, Yu WD*. A prevalence survey of *Wolbachia* in *Polytremis fukia* (*Lepidoptera*: *Hesperiidae*). Journal of Asia-Pacific Entomology. 2016;**19**(2):399-404

[38] Raychoudhury R, Grillenberger BK, Gadau J. Phylogeography of *Nasonia vitripennis* (Hymenoptera) indicates a mitochondrial-*Wolbachia* sweep in North America. Heredity. 2010;**104**:318-326

[39] Stouthamer R, Breeuwer JA, Hurst GD. *Wolbachia pipientis*: Microbial manipulator of arthropod reproduction. Annual Review of Microbiology. 1999;**53**:71-102

[40] Jackel R, Mora D, Dobler S. Evidence for selective sweeps by *Wolbachia* infections: Phylogeny of *Altica* leaf beetles and their reproductive parasites. Molecular Ecology. 2013;**22**:4241-4255

[41] Keller GP, Windsor DM, Saucedo JM, Werren JH. Reproductive effects and geographical distributions of two *Wolbachia* strains infecting the neotropical beetle, *Chelymorpha alternans* Boh. (*Chrysomelidae*, *Cassidinae*). Molecular Ecology. 2004;**13**:2405-2420

[42] Martins C, Souza RF, Bueno OC. Presence and distribution of the endosymbiont *Wolbachia* among *Solenopsis* spp. (*Hymenoptera*: *Formicidae*) from Brazil and its evolutionary history. Journal of Invertebrate Pathology. 2012;**109**:287-296

[43] Shoemaker DD, Dyer KA, Ahrens M, McAbee K, Jaenike J. Decreased diversity but increased substitution rate in host mtDNA as a consequence of *Wolbachia* endosymbiont infection. Genetics. 2004;**168**:2049-2058

[44] Duron O, Bouchon D, Boutin S, Bellamy L, Zhou L, Engelstädter J, Hurst GD. The diversity of reproductive parasites among arthropods: *Wolbachia* do not walk alone. BMC Biology. 2008;**6**:27

[45] Meng XF, Shi M, Chen XX. Population genetic structure of *Chilo suppressalis* (Walker) (*Lepidoptera*: Crambidae): Strong subdivision in China inferred from microsatellite markers and mtDNA gene sequences. Molecular Ecology. 2008;**17**:2880-2897

[46] Zhu ZH, Ye H, Zhang Z. Genetic relationships among four *Bactrocera cucurbitae* geographic populations in Yunnan Province. Chinese Journal of Applied Ecology. 2005;**16**: 1889-1892

[47] Niu CW, Zhang QW, Ye ZH, Luo LZ. Analysis of genetic diversity in different geographic populations of the beet armyworm *Spodoptera exigua* (*Lepidoptera*: *Noctuidae*) with AFLP technique. Acta Entomologica Sinica. 2006;**49**:867-873

[48] Zhang MZ, Kang L. The genetic differentiation among geographic populations of *Locusta migratoria* in China. Scichina. 2005;**35**:220-230

[49] Xu G, Wang GR, Wu KM, Guo YY. Gene flow analysis among different geographical populations of *Helicoverpa armigera* (Hübner) by restriction fragment length polymorphisms. Cotton Science. 2002;**14**:352-355

[50] Yang QC, Zhao HD, Chang Y, Liu L, Zhu JQ, Yu WD, Jiang WB*. Complete mitochondrial genome of the Chinese skipper, *Polytremis jigongi* (*Lepidoptera*: *Hesperiidae*). Mitochondrial DNA Part A. 2016;**27**(4):2375-2376. (SCI)

[51] Jiang WB, Zhu JQ, Zhan L, Chen MH, Song C, Yu WD. Isolation and characterization of microsatellite loci in *Polytremis nascens* (*Lepidoptera*: *Hesperiidae*), and their cross-amplification in related species. Applied Entomology and Zoology. 2014;**49**:177-181

Section 2

Lepidoptera as a Model for Research

Chapter 4

Synergistic Damage Response of the Double-Focus Eyespot in the Hindwing of the Peacock Pansy Butterfly

Mayo Iwasaki and Joji M. Otaki

Additional information is available at the end of the chapter

http://dx.doi.org/10.5772/intechopen.70050

Abstract

Eyespot color patterns in butterfly wings are determined by the putative morphogenic signals from organizers. Previous experiments using physical damage to the forewing eyespots of the peacock pansy butterfly, *Junonia almana* (Linnaeus, 1758), suggested that the morphogenic signals dynamically interact with each other, involving enhancement of activation signals and interactions between activation and inhibitory signals. Here, we focused on the large double-focus fusion eyespot on the hindwing of *J. almana* to test the involvement of the proposed signal interactions. Early damage at a single focus of the prospective double-focus eyespot produced a smaller but circular eyespot, suggesting the existence of synergistic interactions between the signals from two sources. Late damage at a single focus reduced the size of the inner core disk but simultaneously enlarged the outermost black ring. Damage at two nearby sites in the background induced an extensive black area, possibly as a result of the synergistic enhancement of the two induced signals. These results confirmed the previous forewing results and provided further evidence for the long-range and synergistic interactive nature of the morphogenic signals that may be explained by a reaction-diffusion mechanism as a part of the induction model for color-pattern formation in butterfly wings.

Keywords: butterfly wing, color-pattern formation, eyespot, induction model, *Junonia almana*, physical damage, reaction-diffusion model

1. Introduction

Animal bodies often have conspicuous color patterns such as stripes, dots, and eyespots. For example, various color patterns are notable in shells and fishes, and at least some of them have been explained well by some types of reaction-diffusion (RD) models [1–3]. In such models, activation and inhibitory signals interact based on local self-activation and lateral inhibition

[4–6]. A patterning process is initiated randomly, but some of the final outputs, such as zebrafish stripes, are stably constructed. Thus, there is no specific organizer or its associated prepattern that is required to construct the final pattern.

In contrast, eyespot patterns that emerge consistently at particular locations in some fish or other species may require organizers, or something similar, that initiate the determination process at particular locations. Although careful adjustment of boundary conditions for RD equations may be able to computationally specify eyespot locations consistently, such a model may not be robust enough to reproduce a given eyespot pattern in every individual under different environmental and genetic conditions during development. As a compromise, a developmental system that involves both predetermined classical organizers (i.e., sources of the putative morphogenic signals) and RD mechanisms might be more realistic. A potential example of such a system is the spotted mandarin fish, *Synchiropus picturatus* (Peters, 1877), which has many eyespots at distinct and consistent locations [7]. In this fish species, physical damage at the center of the eyespot cannot reduce the eyespot size, and surgical removal of a substantial portion of an eyespot initiates a regeneration process to reconstruct the entire eyespot [7]. These results suggest that the eyespot center does not function as an organizer and that lateral cellular interactions play an important role in constructing an eyespot in this fish species, although an initial specification mechanism of eyespot locations remains enigmatic. Interestingly, ectopic eyespots can be induced by physical damage to the background area between eyespots [7].

Another developmental system that may require both classical organizers and RD mechanisms is the butterfly color-pattern determination system. Butterfly color patterns are constructed based on three major symmetry systems, two peripheral systems, and other accessory systems [8–14]. Each symmetry system is composed of a collection of color-pattern elements. Among these elements, eyespots that belong to the border symmetry system are probably most conspicuous, at least to human eyes, and developmental mechanisms of eyespots have been studied relatively well. The initial specification of the central location of an eyespot has been successfully described by an RD model based on signals from wing veins in developing wing disk [8], although this model may be too fine-tuned to explain the developmental robustness of actual eyespots [15]. Interestingly, the subsequent determination process of an eyespot after the determination of its central location has been explained by a concentration gradient model, a non-RD model [8, 16, 17]. One of the reasons that the butterfly eyespot formation (except for the initial specification) has been considered a non-RD system may be that the center of the prospective eyespot has been known to behave as an organizer, as demonstrated by the following experiments. Cautery-based damage at the center of the prospective eyespot reduces or completely abolishes the prospective eyespot [18, 19], and transplantation of the central cells produces an ectopic eyespot at the transplanted site [18, 20, 21].

However, it has been noted that gradient models cannot explain the extreme diversity of eyespot morphology in nymphalid butterflies [22]. Moreover, gradient models cannot explain the morphological diversity of serial eyespots on the identical wing surface [23]. Furthermore, the status of parafocal elements as a part of the border symmetry system [10, 11, 22, 23] has not been explained by the previous models. Nijhout [15] recently proposed the grass fire model, in which parafocal elements can be produced together with eyespots by a simple RD system. It would not be surprising for the entire process of the butterfly eyespot determination to be solely based on RD mechanisms. However, another way of thinking about the system is that because an RD model in general does not require the existence of organizers, the butterfly color-pattern formation system may be something more than a simple RD system.

Accordingly, a model that includes both an organizer and the essence of an RD system has been proposed, and it is called the induction model [22–24]. For convenience, the induction model can be divided into two stages: the early and late stages. The early stage involves signal expansion and settlement from an organizer, and the late stage involves short-range activation and long-range inhibition, the essence of an RD model. In this model, the activation signals activate themselves, and the activation signals and inhibitory signals interact with each other. When two activation signals from different sources meet, synergistic enhancement may occur.

It is important to stress that the induction model is based on "inductive reasoning," meaning that it is based on collective analysis of many actual butterfly eyespot patterns and physiologically induced color patterns [22, 23, 25]. Thus, the induction model can be applied to "non-typical" distorted eyespots and damage-induced changes, which are not explained well by the gradient models [22, 23]. The induction model is essentially a formal model based on observations, experimental results, and integrative logics, and it is not a computational model that introduces many unknown assumptions. It is true that the induction model proposes unknown mechanisms such as mechanistic waves [13], but these unknowns should be tested and replaced, if necessary, with alternative ideas.

Among the data supporting the induction model is that when a prospective large eyespot is damaged, an adjacent small eyespot becomes larger [26]. This result suggests an inhibitory effect from the prospective large eyespot to the small one. In the induction model, the inhibitory signal is upregulated in the edge of the activation signal, based on the principle of the local (short-range) self-activation and lateral (long-range) inhibition [5]. This inhibition signal works on activation signals not only from its own eyespot but also from other eyespot. Because both activation and inhibition signals behave autonomously once released from organizing cells, the inhibitory signal does not have to affect the signal source to make an eyespot smaller.

Another finding supporting the induction model is that the outermost black ring can be uncoupled from its inner core disk when a prospective eyespot is damaged late [25]. This is also explained by autonomous nature of signals that the induction model proposes. An alternative explanation is that two different chemical morphogens are released. This is not compatible with the conventional gradient model [22], and autonomous behavior of parafocal elements, an equivalent element to "eyespot ring," prefers the induction model [23, 24].

To be sure, this approach is not intended to undermine computational models. Computational models can propose mathematically defined assumption that may be tested systematically, whereas the collective color-pattern analyses were mostly descriptive. However, both approaches are necessary to understand the complexity of butterfly color patterns. A novel and important way to distinguish between the induction model and the gradient model is to examine a fusion eyespot that has two signal sources. A fusion of two eyespots can be explained either by the conventional gradient model or by the induction model. However, the synergistic enhancement by activation signals from two different sources could occur if the induction model (or a similar model) operates (**Figure 1A**). The synergistic enhancement in

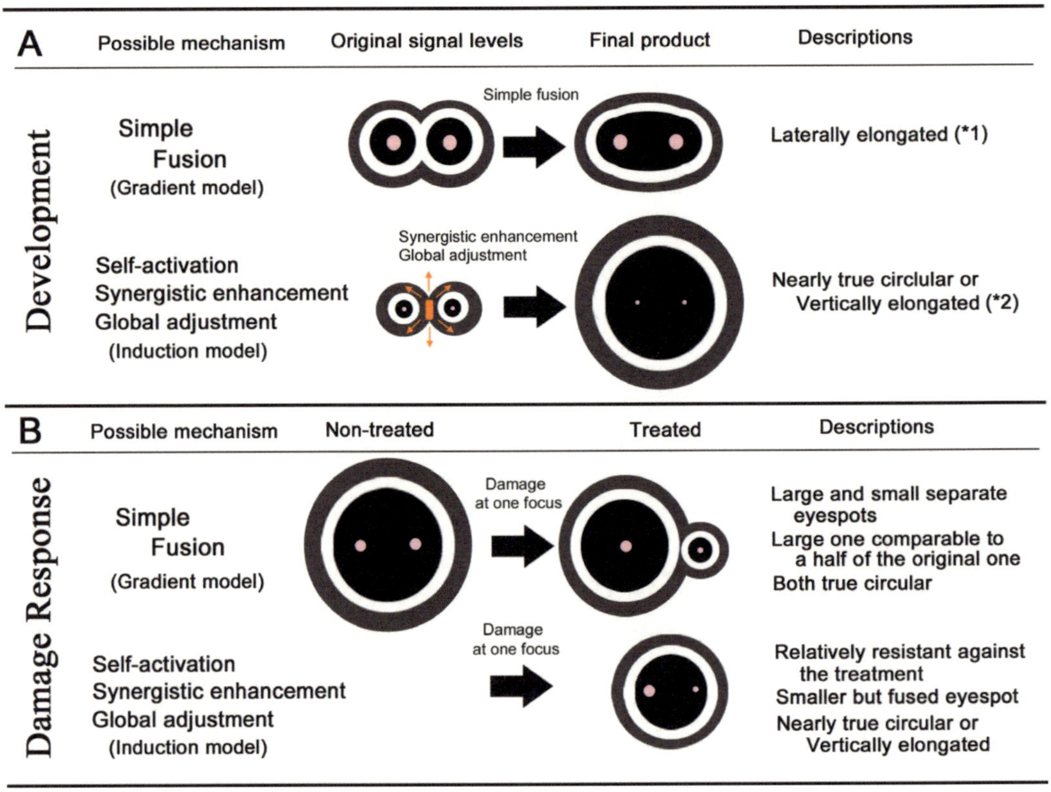

Figure 1. Conceptual distinction between simple fusion (the gradient model) and non-simple fusion (the induction model). Central circular areas indicate organizing centers that emit signals. (A) Distinction of the two modes during development. The simple fusion process produces a laterally elongated eyespot with relatively large numbers of organizing cells (which do not necessarily correspond to the white spots in actual butterflies). In contrast, the non-simple fusion process involves self-activation, synergistic enhancement, and global adjustment and produces a nearly true circular eyespot or vertically elongated eyespot from a relatively small number of organizing cells. The area in which the signals from two sources come into contact (shown in the vertical bar between the two organizing centers) acts as a new "source" of activation signal for the entire eyespot. However, even in the simple fusion, when two sources are relatively closely positioned in comparison with signal levels, the final eyespot may form a nearly true circle (Note *1). Likewise, when self-activation and synergistic enhancement are delayed and inhibited by the emerging inhibitory signals, the final eyespot remains laterally elongated even when the induction model is correct (Note *2). Moreover, signal distribution patterns at the early stage (shown as "Original signal levels") are highly similar between the two models. (B) Distinction of the two modes based on the damage response. If a double-focus eyespot is produced by simple fusion, as predicted by the gradient model, damage at one organizing center of a double-focus eyespot produces separate eyespots, one large and one small. The large one is comparable to half of the original one. Both eyespots are truly circular. In contrast, if a double-focus eyespot is produced by self-activation, synergistic enhancement, and global adjustment, as predicted by the induction model, the eyespot is relatively resistant against the treatment. A small but fused circular or vertically elongated eyespot will result.

http://dx.doi.org/10.5772/intechopen.70050

the induction model can be achieved if activation signals merge together before the upregulation of inhibitory signals around the activation signals. In other words, the final size of a fusion eyespot is determined not by a simple summation of the two independent sources but by a synergistic enhancement process. Importantly, the synergistic enhancement is most active at the boundary between the two sources, and the resultant fused eyespot would thus tend to become nearly a true circle or slightly vertically elongated (**Figure 1A**). In contrast, a simple fusion process will often result in a laterally elongated fused eyespot (**Figure 1A**).

However, it is difficult to distinguish these two mechanistic possibilities simply based on the final morphology of the fusion eyespot alone, given that there may be conditions under which the typical morphology is not attained. For example, when two sources are positioned closely or when two signals are very strong, a simple fusion of the two would produce a near true circle. When the self-activation and synergistic enhancement processes failed to occur for some unexpected kinetic reasons before the upregulation of inhibitory signals, a laterally elongated fusion eyespot may result. Moreover, an essentially indistinguishable morphology will be exhibited by either mechanism at early fusing stages of a pair of eyespots (**Figure 1A**).

Nonetheless, physical damage at a single focus of a double-focus eyespot may resolve these two possibilities. Damage at a single focus would produce two circular eyespots, a large one and a small one, when a single gradient model is operating (**Figure 1B**). In contrast, damage at a single focus would produce a smaller but circular fusion eyespot with two foci if the induction model (or something similar) is operating, because of the synergistic enhancement and the global adjustment of the activation signals (**Figure 1B**). In other words, a double-focus eyespot would behave as if both foci were damaged. Therefore, characterization of the damage response of a double-focus eyespot that is constructed by fusion of two eyespots would test whether the induction model, or something similar, that involves the synergistic signal enhancement is more reasonable than the gradient model.

The best system to test this hypothesis is probably the large dorsal hindwing eyespot of the peacock pansy butterfly, *Junonia almana* (Linnaeus, 1758) (**Figure 2A**). In this paper, this eyespot is called the major eyespot (or the double-focus eyespot), simply because it is large in

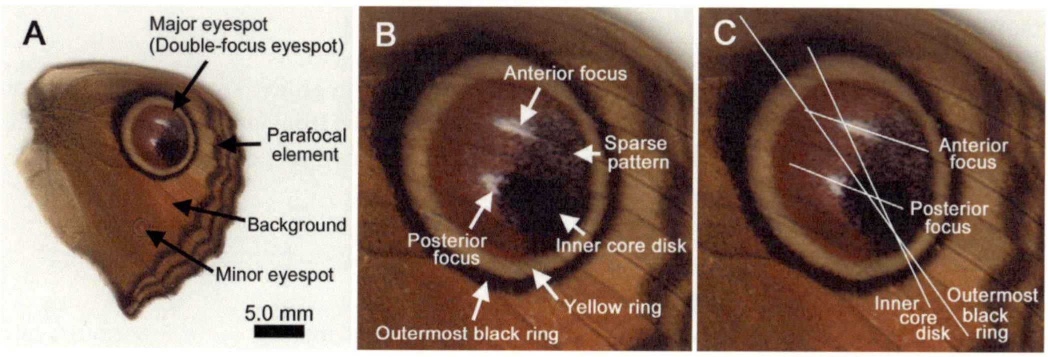

Figure 2. Nomenclature of the hindwing elements and sub-elements of the peacock pansy butterfly, *J. almana*. (A) An entire dorsal hindwing. (B) Higher magnification of the major eyespot and its surroundings. (C) Directions of elongation of sub-elements in the major eyespot. The directions are not in synchrony.

comparison with another eyespot (the minor eyespot) on the same wing surface (**Figure 2A**). Morphologically, the large circular shape with two distantly separate foci of this major eyespot already suggests the feasibility of the induction model (**Figure 1A**). Furthermore, each sub-element (components of an element, some of which are indicated in **Figure 2B**) of the eyespot appears to behave independently; their elongation directions are inconsistent (**Figure 2C**). Such independent behaviors of sub-elements within a given element are called uncoupling [13, 26].

If the large size of this hindwing eyespot is a product of the synergistic enhancement of the signals from two organizers, mechanical damage at a single organizer could reduce the size of the entire eyespot. That is, when one organizer is debilitated by damage, the other intact organizer would "help" to restore the entire eyespot, although small, from the merged center. The eyespot would be relatively resistant to damage because of the synergistic enhancement process. In the line of this argument, it is possible to test whether the prediction of the induction model is consistent with the damage response of the double-focus eyespots.

Here, the damage response of the dorsal hindwing major eyespot of *J. almana* was characterized. Damage to the background was also performed, which has been known to produce an ectopic black spot in the forewing of this species [26] and in other species of nymphalid butterflies [27, 28]. Hindwing damage in butterflies has never been reported except by Nijhout [8]. This is partly because the hindwing is covered by the forewing and is thus invisible from the outside at the pupal stage in butterflies. We have overcome this technical difficulty by directly observing the hindwing development using the forewing-lift method [29, 30]. Nijhout [8] briefly mentioned that the mechanism of eyespot formation in the hindwing may be different from that in the forewing based on the following results from *Junonia coenia* Hübner, 1822: the hindwing eyespot cannot be changed in size by cautery immediately after pupation, whereas the background is still responsive to cautery. This possibility has now also been tested in *J. almana* in this paper.

2. Materials and methods

2.1. Butterflies

The peacock pansy butterfly, *J. almana,* was used throughout this study. This study focused on the hindwing of this butterfly (**Figure 2**). The eggs were collected from females caught from the wild in Ishigaki-jima Island, Okinawa, Japan. Larvae were fed natural host plants in the laboratory at ambient temperature.

2.2. Damage applications and image analysis

After prepupation, pupation time was checked repeatedly at intervals of a few hours, and pupae were categorized into three groups based on time post-pupation: 3–6 h (early), 6–12 h (middle), and 12–18 h (or 12–20 h) (late). Mechanical damage was made at specific positions on the right pupal wings (without a forewing lift) using a stainless needle of 0.50 mm in diameter (Shiga Konchu, Tokyo, Japan). A needle was inserted down to approximately 3 mm in depth and moved up and down five or more times before being removed entirely. The contralateral (left) wing was not damaged because it served as an internal control. The damage sites of the hindwing were determined in advance using a different set of pupae by the forewing-lift method performed in this species [29, 31]. The damaged pupae were kept at ambient temperature until eclosion. The adults that eclosed were frozen immediately after pupation. Wing images were obtained using a Canon MG5730 scanner (Tokyo, Japan). Color-pattern changes of the treated wings were evaluated in reference to the normal color patterns of the non-treated wings of the same individuals.

2.3. Definition of focus

In this paper, an eyespot "focus" was defined as a white spot at the central region of an eyespot in a compartment. The white spots do not necessarily correspond to locations of organizers in this species [31] and also in other species [32]. However, because a white spot indicates an approximate location of an organizer in this species, the white spot is conventionally called the focus in this paper.

3. Results

3.1. Anterior damage to the major eyespot

The anterior focus of the major eyespot was damaged at 3–6 h post-pupation ($n = 15$). In 12 out of 15 cases (80%), the major eyespot was reduced in overall size (**Figure 3A–C**). Importantly, the entire eyespot (not only the anterior side but also the posterior side) decreased in size in most cases, although the damage was placed only at the anterior focus, suggesting dynamic interactions between the anterior and posterior signals during development to determine the final size and shape of the major eyespot. One individual exhibited minor size reduction at the anterior side but not clearly at the posterior side (**Figure 3C**). In 3 out of 15 cases (20%), the reduction was not clear. The failure of the size reduction was probably because the damage was mistakenly (but unavoidably) placed at the semi-focal point. In these semi-focal damage samples, coloration inside the core disk was disrupted, an ectopic yellow area emerged, and the anterior focus (white spot) was elongated toward the damage site (**Figure 3D**).

Similarly, the anterior focus of the major eyespot was damaged, but much later, at 12–20 h post-pupation ($n = 4$). In all 4 cases (100%), the outer black ring was enlarged in all directions, but the inner core disk was reduced in size, although to a small degree (**Figure 3E**), showing an uncoupling response between these two sub-elements.

3.2. Posterior damage to the major eyespot

The posterior focus of the major eyespot was damaged at 3–6 h post-pupation ($n = 9$). In 2 out of 9 cases (22%), the overall double-focus eyespot was moderately reduced in size (**Figure 3F**). The overall shape remained circular. No effect was observed in 7 out of 9 cases (78%), indicating the relatively low sensitivity of the posterior focus to damage in comparison with the anterior focus.

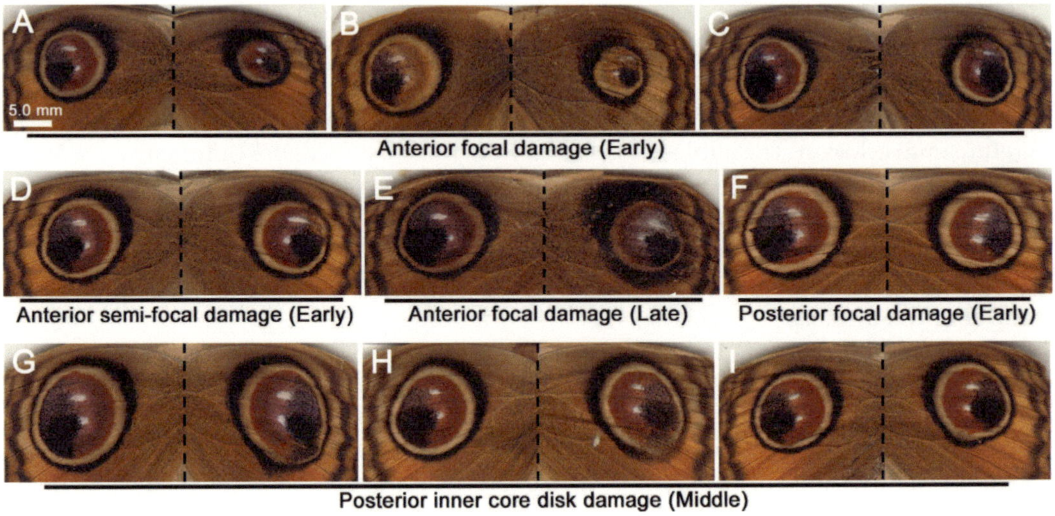

Figure 3. Damage-induced color-pattern changes in the major eyespot of the hindwing of *J. almana*. (A and B) Early damage at the anterior focus. In these typical cases, the entire eyespot was reduced in size, but it was still circular. (C) Early damage at the anterior focus. In this exceptional case, only the treated anterior side was clearly reduced, although the reduction level was minor. (D) Early damage at the anterior semi-focal site. A yellow area emerged inside the inner core disk. The anterior focus (white dot) was elongated toward the damage site. (E) Late damage at the anterior focus. The outermost black ring expanded in all directions, but the inner core disk was reduced in size. (F) Early damage at the posterior focus. The entire eyespot size was reduced slightly in size. (G-I) Middle (mid-term) damage at the posterior inner core disk. Response was largely local. Difference from the anterior focal damage shown in E is notable.

Then, the core disk was damaged in the posterior side, avoiding the posterior focus, at 6–12 h post-pupation ($n = 18$). In 10 out of 18 individuals (56%), the treatment induced the formation of a yellow area inside the inner core disk and made the coloration boundaries fuzzy (**Figure 3G** and **H**). Even in these cases, the overall size of the double-focus eyespot did not change. That is, changes were restricted to the immediate vicinity of the damage site. No or very minor effect was observed in 8 out of 18 cases (44%) (**Figure 3I**).

3.3. Damage to the outermost black ring of the major eyespot or in its close vicinity

The outermost black ring of the major eyespot was damaged at 6–12 h post-pupation. Because eyespot size and shape were slightly different from individual to individual, damage was made without distinction at the outermost black ring, at the yellow ring, or at the background immediately close to the outermost black ring ($n = 73$). Among these 73 treated individuals, a small black dot with a yellow area inside ($n = 21$; 29%) or without a yellow area ($n = 17$; 23%) emerged in a close proximity to the major eyespot. In another set of individuals, such a dot emerged on the yellow ring that accompanied the extrusion of the outermost black ring ($n = 10$; 14%) (**Figure 4A**). In more extensive cases ($n = 13$; 18%), entire eyespot shape was disrupted, with the extrusion of the inner core disk toward the damage site (**Figure 4B** and **C**), with a broken outermost ring (**Figure 4D**), or with the enlargement of the yellow area (**Figure 4E**). In many of these cases, both the outermost black ring and the core disk were distorted toward the damage site.

http://dx.doi.org/10.5772/intechopen.70050

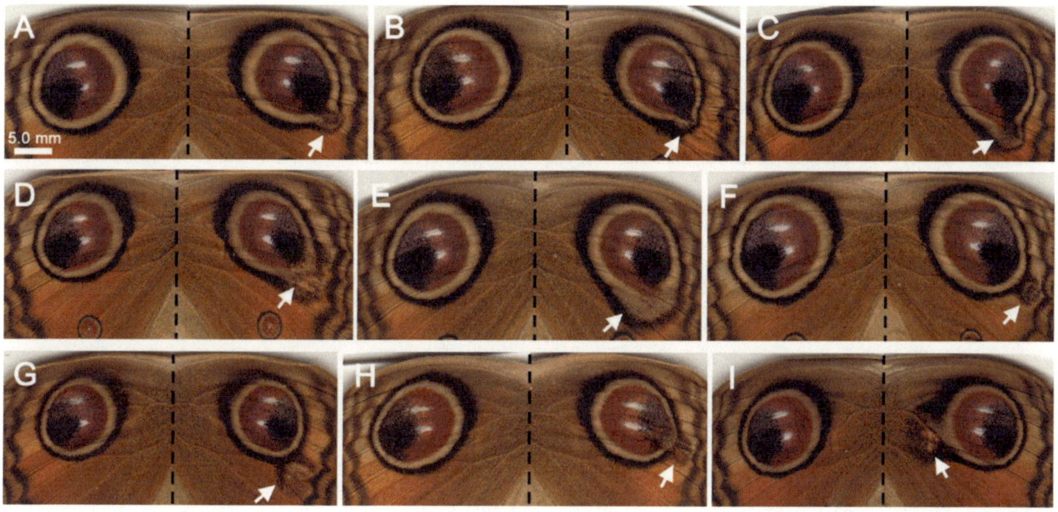

Figure 4. Damage at or around the outermost black ring of the major eyespot in the hindwing of *J. almana*. Damaged points are indicated by arrows. (A) Small ectopic black ring on the yellow ring. (B and C) Extrusion of the inner core disk and other structures. (D) Breaking of the outermost black ring. (E) Expansion of the black ring and yellow area. (F) Small ectopic arc and ring that fuse with the outermost black ring of the major eyespot but not with parafocal element. (G) Black arc (and vague black area inside) that fuses with the outermost black ring but not with parafocal elements. (H) Induced black area (arrow) and a sparse pattern that is induced in the posterior side of the inner core disk. A sparse pattern is present only in the anterior side of the inner core disk in non-treated individuals. (I) Large black area with an orange area inside induced by damage at the proximal side of the major eyespot.

In a different set of individuals ($n = 11$; 15%), a double black ring or a similar feature emerged in the background area immediately facing the eyespot (**Figure 4F** and **G**). Interestingly, in these cases, the double ring structure was not expressed clearly in the side facing the parafocal elements. In one case ($n = 1$; 1%), the damage-induced structure in the background was very minor, but a sparse pattern was induced in the posterior side of the black core disk (**Figure 4H**).

In addition to damage at the distal side of the major eyespot, the proximal side of the major eyespot was damaged at the outermost black ring at 12–18 h post-pupation ($n = 21$) (**Figure 4I**). Among these 21 treated individuals, an ectopic yellow region that was surrounded by a black area was produced in most individuals ($n = 18$; 86%). This induced black area fused with the outermost black ring of the eyespot smoothly, and the entire eyespot was distorted only slightly, if at all, toward the damage site. No effect was observed in 3 out of 21 cases (14%).

3.4. Damage to the background between the major and minor eyespots

To understand the reactivity of the background, the background between the major and minor eyespots was damaged at 12–18 h post-pupation ($n = 9$). A black area was induced in all treated individuals (100%) (**Figure 5A**). In the most severe cases, a yellow area emerged at the center (**Figure 5B**). In the individual shown in **Figure 5B**, the induced black area fused with the outermost black rings of the major and minor eyespots, and the minor eyespot was distorted toward the damage site, where a yellow area emerged inside the induced black area.

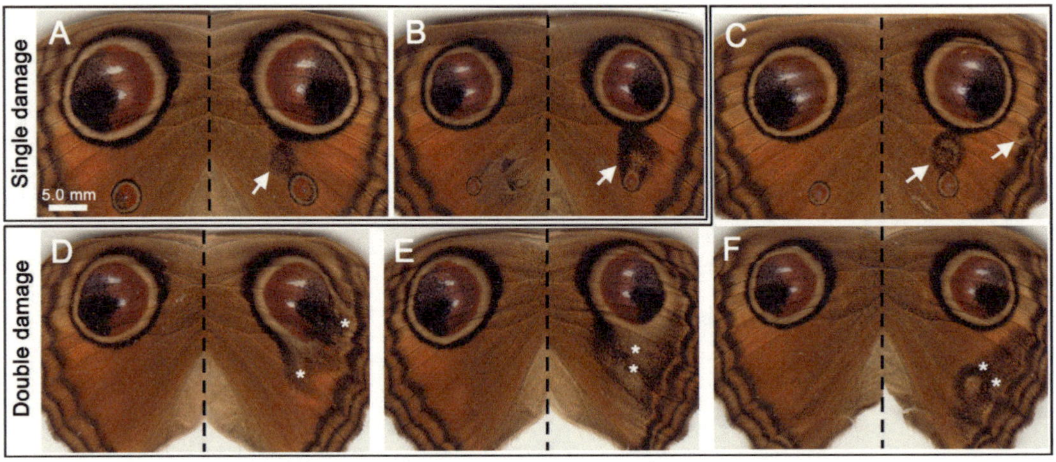

Figure 5. Damage to the background in the hindwing of *J. almana*. Damaged points are indicated by arrows or asterisks. (A and B) Single point of damage between the major and minor eyespots. A black area was induced. (C) Double damage at the area between the major and minor eyespots and near parafocal elements. A double ring structure and a small dot were induced. The two damage sites responded independently. (D) Double damage at the area between the major and minor eyespots and near parafocal elements. The major eyespot was ruptured. (E) Double damage between the major and minor eyespots, resulting in extensive modifications. (F) Double damage near the minor eyespot.

3.5. Double background damage

In the experiments described above, a single site per individual was damaged. Here, to understand possible interactions between damage-induced signals, two sites in the background were damaged. First, two distant sites in the background were damaged, one between the major and minor eyespots and one between the major eyespot and parafocal elements, at 12–18 h post-pupation (n = 6). In all 6 cases (100%), a black area was induced at both damage sites without any noticeable interaction. In one of these cases, the proximal site (an area between the major and minor eyespots) was more extensive than the distal site (an area near parafocal elements) (**Figure 5C**). Indeed, in this individual, the induced black spot at the distal site was very close to parafocal elements but did not fuse with them. Instead, the ectopic small black spot appeared to "repel" the nearby parafocal element. In 2 out of 6 cases (33%), including the individual shown in **Figure 5C**, a clear double ring appeared at the proximal site. In one extensive case, the major eyespot opened up with the extensive light black area (**Figure 5D**). In this individual, the induced black area again did not fuse with parafocal elements; there was a clear gap between them.

Then, two closely positioned sites in a wing between the major and minor eyespots were damaged at 12–18 h post-pupation (n = 5). All treated individuals (100%) showed marked disruption of the major eyespot. The outermost black ring was ruptured, and the induced black area covered extensive portions of the background (**Figure 5E**). However, the induced black area again could not invade parafocal elements. There was a narrow but distinct gap between the induced black area and parafocal elements.

When two closely positioned sites around the minor eyespots were damaged at 12–18 h post-pupation (n = 15), 14 out of 15 cases (93%) showed induction of a black area. One of them showed an extensive black area along the parafocal elements (**Figure 5F**). Although the ectopic black area fused smoothly with the outermost black ring of the major eyespot, there was no fusion between the ectopic black area and parafocal elements. Interestingly, the distinct minor eyespot was not present in the normal wing in this particular individual, but the induced black area did not enter the minor eyespot area, thereby demonstrating the existence of the imaginary ring at that site. No effect was observed in one case (7%).

4. Discussion

4.1. Hindwing eyespot response

In the present study, response profiles of the double-focus eyespot and its surrounding wing surface in the hindwing of *J. almana* were obtained. This eyespot on the dorsal hindwing is quite large, and simply because it has two foci in two compartments, one can discern that this eyespot is a fusion of two original eyespots. The response profile of the hindwing double-focus eyespot that was obtained in the present study is largely consistent with that of the forewing single-focus eyespot of *J. almana* reported previously [26]. It was confirmed that focal damage, but not non-focal damage, dramatically changed the overall eyespot size, supporting the idea that the focal cells function as organizing cells. However, it was found that the posterior focus was less sensitive than the anterior focus. This sensitivity difference probably reflects different developmental periods when organizing cells are active. The insensitivity of the hindwing eyespot to cautery-based damage in *J. coenia* [8] may simply be attributable to earlier species-specific differentiation of organizing cells, before the treatment time point. In that case, the insensitivity does not suggest any fundamental mechanistic difference between the forewing and hindwing.

4.2. Synergistic response to focal damage

The double-focus eyespot of *J. almana* provided an ideal system to test the dynamic interactions between the signals from two different sources. Importantly, the early anterior focal damage made the entire double-focus eyespot small as if both foci were damaged, and the treated eyespot kept its circular shape. These results suggest the existence of dynamic interaction, possibly synergistic enhancement, of the activation signals from two sites, confirming the feasibility of the induction model over the conventional gradient model. In one case, a minor size reduction only at the treated anterior side was observed. However, simply because the change was minor (likely due to incomplete damage), this case does not support the gradient model.

Additionally, the late anterior focal damage enlarged the outermost black ring but reduced the size of the inner core disk. The enhancement of the outermost black ring was not restricted to the anterior side; the enlargement was in all directions. This uncoupling behavior between the outermost black ring and the inner core disk within the same eyespot is indeed consistent with the late damage results of the forewing eyespot [26]. This uncoupling response can be explained if the normal signals are wave-like (which means that the signals behave independently from their source once released) and if the induced signal was added to the normal signal for the outermost black ring. The normal signal for the inner core disk was being released at the time of damage, and because some of the organizing cells were destroyed by physical damage, the inner core disk became smaller. This uncoupling response is an indication of independence of the signal for each sub-element (i.e., the outer black ring and the inner core disk). The wave-like nature of signals is also highlighted in these results. These results are not explainable by the gradient model.

4.3. Response to other types of damage

Semi-focal damage at the anterior side produced a yellow area inside the inner core disk, which is also consistent with the forewing results [26]. This result is also difficult to explain using a gradient model. A threshold increase in response to damage may be a remedy, but a threshold decrease should also be introduced to explain the induced black area in the background. Furthermore, the induced double-ring structures in the background require multiple threshold sets to be explained by the gradient model. These damage-induced rings have been shown to have scale structures that are similar to those of normal eyespots [31]. These complicated threshold arrangements are too complex to accept as a theoretical framework for color-pattern determination in butterfly wings.

Interestingly, the white focal area was elongated toward the damage site in the semi-focal damage. Notably, developmental signals for the white "focal" spot and the eyespot body to which that white spot belongs do not have to be identical [31, 32]. Indeed, the white focal spot is likely uncoupled from the rest of the sub-elements [32].

Damage at or around the outer black ring produced various results. A small black ring was produced in the yellow ring in some cases, but in other cases, the inner core disk, the yellow ring, and the outer black ring were often "pinched off" from the normal shape of the eyespot, suggesting that the ectopically induced signals are able to merge with natural signals locally. In other words, spontaneous and artificially induced signals are indistinguishable to developing scale cells. Furthermore, the extrusion of both the outermost black ring and the inner core disk toward a damaged site suggests that serial lateral interactions keep their shapes, which is reminiscent of the eyespots of the spotted mandarin fish [7]. In addition to these local effects, overall shape changes of the treated eyespots were often observed, although not extensively in response to this manipulation.

4.4. Synergistic response to double background damage

Double-damage experiments that produced extensive black areas confirmed that the induced signals at two sites can be combined to produce strong effects. It is likely that when two sites of damage were close enough, the induced area was more than a simple summation of two areas induced independently by two single damage treatments. These results can be interpreted as evidence for synergistic enhancement of two artificially upregulated signals in the hindwing of this species. This synergistic enhancement process may also occur spontaneously in the double-focus eyespot during development.

In some cases, the black signals induced by double damage merged with the outermost black ring of the natural eyespot, resulting in the rupture of the eyespot. This result again demonstrates the indistinguishability of the natural and induced signals. Somewhat surprisingly, the large black area highlighted one of the "inhibitory areas" that are usually invisible but associated with parafocal elements and the minor eyespot. That is, in an individual lacking the distinct minor eyespot, the induced black area could not invade the area surrounding the minor eyespot. This invisible area might have arisen if the inhibitory signal became stronger and larger in that area than the activation signal for black areas during development of the minor eyespot. Similarly, the induced black signal could not make contact with parafocal elements, suggesting that the inhibitory signals are present along parafocal elements. The similar inhibitory area that surrounds an eyespot has generally been termed the imaginary ring [13, 22]. The reason that the imaginary ring was not observed around the major eyespot is not well understood, but development of an imaginary ring around the major eyespot may require additional time before the end of the pattern determination period.

4.5. Possible mechanisms

Overall, these results strongly suggest that the signals that determine the final scale color of a given scale cell are highly dynamic. It is likely that a reaction-diffusion mechanism, as a part of the induction model, operates in the butterfly color-pattern determination system. The induction model consists of two stages. The early stage is a dissipation of signals from their source, and the late stage is essentially a reaction-diffusion mechanism that involves short-range activation and long-range inhibition. It is speculated that calcium signals play an important role in color-pattern determination in the late stage of the induction model; calcium signals traveling on the developing pupal wings have been observed [33]. On the basis of a linear relationship between scale size and cell size [34–36] and a relationship between scale color and scale size [31], it has been proposed that the putative morphogenic signals from organizers are ploidy signals that determine cellular size via polyploidization [31]. Calcium signals may play an important role in polyploidization.

The early stage of the induction model proposes that a signal moving slowly from its source is the original morphogen that subsequently triggers calcium waves as an activation signal. It is speculated that this slow-moving signal is waves of mechanical distortion [25]. Importantly, organizing centers are present as physical bumps or indentations [32, 37]. These organizing centers can be identified as the pupal cuticle spots in pupae [38, 39]. Based on this fact and other observations, the distortion hypothesis has been proposed, in which physical distortion of the wing tissue functions as the primary morphogenic signal [13]. Physical distortion of the wing epithelial sheet will be created when cells at the organizers selectively increase their sizes via an increase in cell number or polyploidization. Ecdysone receptor is expressed in focal cells in *J. coenia* [40], and such an increase of cells has been observed in eyespot centers of *Bicyclus anynana* (Butler, 1879) in response to ecdysone; this increase likely results in larger eyespots [41]. The finding that ecdysone injection into pupae of *J. almana* does not affect eyespot size but does change background coloration [42] is to be reconciled with the observation that ecdysone receptor is responsible for an increase in organizing cells in *B. anynana*.

In physical damage experiments, mechanical distortion of the wing epithelial sheet is probably introduced, nicely mimicking the natural developmental process that involves physical distortion waves. Although there are some classical histological studies on developing wing tissues of butterflies and moths [34, 43, 44], real-time live imaging studies have just begun on developing wing tissues and organizing cells [30, 37, 45]. The distortion hypothesis should be tested in the future in light of the importance of mechanical forces in development [46, 47]. Compatibility of these proposed mechanisms with other related mathematical models for eyespot focus determination [48–50] is also to be investigated in the future.

5. Conclusions

The present study provided experimental evidence that morphogenic signals for eyespot color patterns are able to synergistically interact with each other, focusing on the damage-induced color-pattern changes of the double-focus eyespot in the hindwing of the peacock pansy butterfly, *J. almana*. The present results may be explained by a reaction-diffusion mechanism as a part of the induction model, but not by the conventional gradient model. A different set of experiments that removed the surface contact from the posterior side of the hindwing major eyespot results in miniaturization of both the anterior and posterior sides of the eyespot [51], suggesting synergistic interactions between the two focal signals that are consistent with the present study.

Acknowledgements

The authors greatly thank Masaki Iwata and other members of the BCPH Unit of Molecular Physiology for technical help and discussion. This project was supported by the basic research fund from University of the Ryukyus and by JSPS KAKENHI, Grant-in-Aid for Scientific Research (C), Grant number 16K07425. MI conducted the experiments and analyzed the data. JMO designed and conducted the experiments, analyzed the data, and wrote the paper. The authors declare that they have no competing interests.

Author details

Mayo Iwasaki and Joji M. Otaki*

*Address all correspondence to: otaki@sci.u-ryukyu.ac.jp

The BCPH Unit of Molecular Physiology, Department of Chemistry, Biology and Marine Science, Faculty of Science, University of the Ryukyus, Nishihara, Okinawa, Japan

References

[1] Kondo S, Asai R. A reaction-diffusion wave on the skin of the marine angelfish *Pomacanthus*. Nature. 1995;**376**:765-768

[2] Meinhardt H. The Algorithmic Beauty of Sea Shells. 4th ed. Berlin: Springer-Verlag; 2009. p. 269

[3] Watanabe M, Kondo S. Is pigment patterning in fish skin determined by the Turing mechanism? Trends in Genetics. 2015;**31**:88-96

[4] Gierer A, Meinhardt H. A theory of biological pattern formation. Kybernetik. 1972; **12**:30-39

[5] Meinhardt H, Gierer A. Pattern formation by local self-activation and lateral inhibition. BioEssays. 2000;**22**:753-760

[6] Kondo S, Miura T. Reaction-diffusion model as a framework for understanding biological pattern formation. Science. 2010;**329**:1616-1620

[7] Ohno Y, Otaki JM. Eyespot colour pattern determination by serial induction in fish: Mechanistic convergence with butterfly eyespots. Scientific Reports. 2012;**2**:290

[8] Nijhout HF. The Development and Evolution of Butterfly Wing Patterns. Washington: Smithsonian Institution Press; 1991. p. 297

[9] Nijhout HF. Elements of butterfly wing patterns. Journal of Experimental Zoology. 2001;**291**:213-225

[10] Dhungel B, Otaki JM. Local pharmacological effects of tungstate on the color-pattern determination of butterfly wings: A possible relationship between the eyespot and parafocal element. Zoological Science. 2009;**26**:758-764

[11] Otaki JM. Colour-pattern analysis of parafocal elements in butterfly wings. Entomological Science. 2009;**12**:74-83

[12] Otaki JM. Color pattern analysis of nymphalid butterfly wings: Revision of the nymphalid groundplan. Zoological Science. 2012;**29**:568-576

[13] Otaki JM. Self-similarity, distortion waves, and the essence of morphogenesis: A generalized view of color pattern formation in butterfly wings. In: Sekimura T, Nijhout HF, editors. Integrative Approach to the Analysis of the Diversity and Evolution of Butterfly Wing Patterns. Singapore: Springer Nature Singapore; 2017. pp. 119-152

[14] Taira W, Kinjo S, Otaki JM. The marginal band system in nymphalid butterfly wings. Zoological Science. 2015;**32**:38-46

[15] Nijhout HF. The common developmental origin of eyespot and parafocal elements; and a new model-mechanism for color pattern formation. In: Sekimura T, Nijhout HF, editors. Integrative Approach to the Analysis of the Diversity and Evolution of Butterfly Wing Patterns. Singapore: Springer Nature Singapore; 2017. pp. 3-19

[16] Nijhout HF. The color patterns of butterflies and moths. Scientific American. 1981;**245**: 104-115

[17] Monteiro A, French V, Smit G, Brakefield PM, Metz JA. Butterfly eyespot patterns: Evidence for specification by a morphogen diffusion gradient. Acta Biotheoretica. 2001;**49**:77-88

[18] Nijhout HF. Pattern formation on lepidopteran wings: Determination of an eyespot. Developmental Biology. 1980;**80**:267-274

[19] French V, Brakefield PM. The development of eyespot patterns on butterfly wings: Morphogen sources or sink? Development. 1992;**116**:103-109

[20] French V, Brakefield PM. Eyespot development on butterfly wings: The focal signal. Developmental Biology. 1995;**168**:112-123

[21] Brakefield PM, Gates J, Keys D, Kesbeke F, Wijngaarden PJ, Monteiro A, French V, Carroll SB. Development, plasticity and evolution of butterfly eyespot patterns. Nature. 1996;**384**:236-242

[22] Otaki JM. Color-pattern analysis of eyespots in butterfly wings: A critical examination of morphogen gradient models. Zoological Science. 2011;**28**:817-827

[23] Otaki JM. Generation of butterfly wing eyespot patterns: A model for morphological determination of eyespot and parafocal element. Zoological Science. 2011;**28**:817-827

[24] Otaki JM. Structural analysis of eyespots: Dynamics of morphogenic signals that govern elemental positions in butterfly wings. BMC Systems Biology. 2012;**6**:17

[25] Otaki JM. Physiologically induced color-pattern changes in butterfly wings: Mechanistic and evolutionary implications. Journal of Insect Physiology. 2008;**54**:1099-1112

[26] Otaki JM. Artificially induced changes of butterfly wing colour patterns: Dynamic signal interactions in eyespot development. Scientific Reports. 2011;**1**:111

[27] Nijhout HF. Cautery-induced color patterns in *Precis coenia* (Lepidoptera: Nymphalidae). Journal of Embryology and Experimental Morphology. 1985;**86**:191-203

[28] Brakefield PM, French V. Eyespot development on butterfly wings: The epidermal response to damage. Developmental Biology. 1995;**168**:98-111

[29] Kusaba K, Otaki JM. Positional dependence of scale size and shape in butterfly wings: Wing-wide phenotypic coordination of color-pattern elements and background. Journal of Insect Physiology. 2009;**55**:175-183

[30] Iwata M, Ohno Y, Otaki JM. Real-time *in vivo* imaging of butterfly wing development: Revealing the cellular dynamics of the pupal wing tissue. PLoS One. 2014;**9**:e89500

[31] Iwata M, Otaki JM. Spatial patterns of correlated scale size and scale color in relation to color pattern elements in butterfly wings. Journal of Insect Physiology. 2016;**85**:32-45

[32] Iwata M, Otaki JM. Focusing on butterfly eyespot focus: Uncoupling of white spots from eyespot bodies in nymphalid butterflies. SpringerPlus. 2016;**5**:1287

[33] Ohno Y, Otaki JM. Spontaneous long-range calcium waves in developing butterfly wings. BMC Developmental Biology. 2015;**15**:17

[34] Henke K. Ueber die verschiendenen Zellteilungsvorgänge in der Entwicklung des beschuppten Flügelepithelis der Mehlmotte *Ephestina kühniella* Z. Biologisches Zentralblatt. 1946;**65**:120-135

[35] Sondhi KH. The biological foundations of animal patterns. The Quarterly Review of Biology. 1963;**38**:289-327

[36] Cho EH, Nijhout HF. Development of polyploidy of scale-building cells in the wings of *Manduca sexta*. Arthropod Structure and Development. 2012;**42**:37-46

[37] Iwasaki M, Ohno Y, Otaki JM. Butterfly eyespot organiser: *In vivo* imaging of the prospective focal cells in pupal wing tissues. Scientific Reports. 2017;**7**:40705

[38] Otaki JM, Ogasawara T, Yamamoto H. Morphological comparison of pupal wing cuticle patterns in butterflies. Zoological Science. 2005;**22**:21-34

[39] Taira W, Otaki JM. Butterfly wings are three-dimensional: Pupal cuticle focal spots and their associated structures in *Junonia* butterflies. PLoS One. 2016;**11**:e0146348

[40] Koch PB, Merk R, Reinhardt R, Weber P. Localization of ecdysone receptor protein during colour pattern formation in wings of the butterfly *Precis coenia* (Lepidoptera: Nymphalidae) and co-expression with Distal-less protein. Development Genes and Evolution. 2003;**212**:571-584

[41] Monteiro A, Tong X, Bear A, Liew SF, Bhardwaj S, Wasik BR, Dinwiddie A, Bastianelli C, Cheong WF, Wenk MR, Cao H, Prudic KL. Differential expression of ecdysone receptor leads to variation in phenotypic plasticity across serial homologs. PLoS Genetics. 2015;**11**:e1005529

[42] Otaki JM. Reversed type of color-pattern modifications of butterfly wings: A physiological mechanism of wing-wide color-pattern determination. Journal of Insect Physiology. 2007;**53**:526-537

[43] Mayer AG. The development of the wing scales and their pigment in butterflies and moths. Bulletin of the Museum of Comparative Zoölogy at Harvard College. 1896;**XXIX**(5):219-236

[44] Nardi J, Magee-Adams SM. Formation of scale spacing patterns in a moth wing: I. Epithelial feet may mediate cell rearrangement. Developmental Biology. 1986;**116**:278-290

[45] Ohno Y, Otaki JM. Live cell imaging of butterfly pupal and larval wings *in vivo*. PLoS One. 2015;**10**:e0128332

[46] Iskratsch T, Wolfenson H, Sheetz MP. Appreciating force and shape—the rise of mechanotransduction in cell biology. Nature Reviews Molecular Cell Biology. 2014;**15**:825-833

[47] Gilmour D, Rembold M, Leptin M. From morphogen to morphogenesis and back. Nature. 2017;**541**:311-320

[48] Sekimura T, Venkataraman C, Madzvamuse A. A model for selection of eyespots on butterfly wings. PLoS One. 2015;**10**:e0141434

[49] Evans TM, Marcus JM. A simulation study of the genetic regulatory hierarchy for butterfly eyespot focus determination. Evolution and Development. 2006;**8**:273-283

[50] Marcus JM, Evans TM. A simulation study of mutations in the genetic regulatory hierarchy for butterfly eyespot focus determination. BioSystems. 2008;**93**:250-255

[51] Otaki JM. Contact-mediated eyespot color-pattern changes in the peacock pansy butterfly: Contributions of mechanical force and extracellular matrix to morphogenic signal propagation. In: Perveen FK, editor. Lepidoptera. Rijeka, Croatia: InTech; 2017

Chapter 5

Contact-Mediated Eyespot Color-Pattern Changes in the Peacock Pansy Butterfly: Contributions of Mechanical Force and Extracellular Matrix to Morphogenic Signal Propagation

Joji M. Otaki

Additional information is available at the end of the chapter

http://dx.doi.org/10.5772/intechopen.70098

Abstract

Butterfly wing color patterns are developmentally determined by morphogenic signals from organizers in the early pupal stage. However, the precise mechanism of color-pattern determination remains elusive. Here, mechanical and surface disturbances were applied to the pupal hindwing of the peacock pansy butterfly *Junonia almana* (Linnaeus, 1758) to examine their effects on color-pattern determination. Using the forewing-lift method immediately after pupation, a small stainless ball was placed on the prospective major eyespot or background of the developing dorsal hindwing to cause a wing epithelial distortion, resulting in deformation of the major eyespot. When the exposed dorsal hindwing was covered with a piece of plastic film or placed on a surface of a glass slide, an adhesive tape, or a silicone-coated glassine paper, the major eyespot was effectively reduced in size without a direct contact with the covering materials. The latter two treatments additionally induced the size reduction of the minor eyespot and proximal displacement and broadening of parafocal elements through a direct contact, being reminiscent of the temperature-shock-type modifications. These results suggest the importance of mechanical force and physicochemical properties of planar epithelial contact surface (i.e., extracellular matrix) to propagate morphogenic signals for color-pattern determination in butterfly wings.

Keywords: butterfly wing, color-pattern formation, distortion hypothesis, eyespot, induction model, mechanical distortion, morphogen

1. Introduction

In any biological systems, cells are placed in an environment where not only chemical information but also mechanical information change over time. The biologically relevant chemical and mechanical information is to be extracted by cells in real time. Chemical information is obtained via receptor molecules that are often specific to soluble chemicals such as hormones, cytokines, growth factors, neurotransmitters, and morphogens. Mechanical information is obtained via integrins and other membrane-spanning molecules that connect the extracellular matrix molecules with the intracellular actomyosin filaments [1]. In this sense, physicochemical properties of the extracellular matrix contribute to information signaling. At the organismal level, chemical information and mechanical information are obtained through the olfactory and gustatory systems and the mechanosensory system, respectively. Because both "modalities" are necessary for any cellular systems, immature cells may take advantage of both modalities to "sense" their environment to determine their own fate for differentiation during development.

Morphogenesis is sequential processes that involve three-dimensional changes of epithelial sheets [2, 3]. In other words, mechanical changes are necessarily involved during morphogenesis. However, a conventional understanding of the developmental fate determination process almost exclusively focuses on chemical signals and their reception, which is manifested, for example, as the gradient model for positional information [4, 5]. By contrast, mechanical signals and their reception have not been acknowledged well in developmental biology. Recent advancement of mechanobiology [1] will help to understand mechanical aspects of cells and tissues during development. However, a pattern formation system that relies on mechanical aspects of tissues has not been investigated sufficiently yet.

Butterfly wings exhibit extreme diversity of color patterns based on developmental and evolutionary modifications of the nymphalid groundplan [6–11]. The butterfly wing system is largely a two-dimensional entity as depicted in the nymphalid groundplan, but strictly speaking, it is three-dimensional; organizers for color patterns are located at the bottom (or top) of an indentation (or a bump) of the wing epithelium in the pupal stage, and this epithelial structure is reflected as pupal cuticle spots [12–14]. Furthermore, this three-dimensionality is reflected in adult wings [13]. Considering these facts, the distortion hypothesis has been proposed, in which mechanical waves generated by oscillatory physical disturbances of the wing epithelial tissue behave as morphogenic (morphogen-like) signals [3].

In this study, the possibility that mechanical and physicochemical properties of extracellular milieu of the epithelial tissue play an important role in morphogenic signal propagation was explored. It has been suggested that some extracellularly secreted molecules such as the Wnt family and TGF-β family proteins behave as chemical morphogens for color-pattern determination in butterflies [15, 16], although how and where these chemical morphogens are distributed are not known. Furthermore, other molecules that could regulate color patterns such as a transcription factor Distal-less have been studied in butterfly wings [17–20]. These molecular signals and regulators are certainly important and compatible with mechanical signal transduction; in a recent model, gene expression regulations are elicited in response to mechanical signals [3].

Here, this study concentrates on the dorsal hindwing of the peacock pansy butterfly, *Junonia almana* (Linnaeus, 1758). This butterfly has a large double-focus eyespot on the dorsal side of the hindwing. This eyespot is a fusion of the two original eyespots, but it is called the major eyespot as a singular entity. The dorsal hindwing also has a much smaller eyespot called the minor eyespot, which is sometimes nonexistent, and the parafocal elements, which, together with eyespots, belong to the border symmetry system. Importantly, the background area has a light orange coloration and does not harbor anything like semi-element or pseudo-element, which probably exists in the blue pansy butterfly, *J. orithya* (Linnaeus, 1758) [21–23]. Furthermore, several versions of color-pattern modifications in response to various treatments have already been known in the peacock pansy butterfly; it has been used for the injections of sodium tungstate, and ecdysteroid and for temperature shock [24] and for physical damage [25, 26]. The scale-size and scale-color distributions have also been recorded in detail in this species [27].

In the present study, the forewing-lift operation was employed, which has been developed and used for several experiments [8, 18, 21, 27–30]. This operation made it possible to insert a small stainless ball between the forewing and the hindwing to disturb the planar epithelial surface. Furthermore, the operation made it possible to cover the hindwing surface with various covering materials. It is likely that the hindwing surface is covered only with a thin layer, if any, of cuticle. This means that the cellular environment of the extracellular matrix can be manipulated directly. Here, various color-pattern modifications were successfully obtained, including the high-level size reduction of the major eyespot, on the dorsal hindwing by the forewing-lift method using small stainless balls and various covering materials. Importantly, modifications of the minor eyespot and parafocal elements were also obtained, which were reminiscent of the temperature-shock-type (TS-type) modifications known in this species [24].

These results highlight the importance of mechanical force and extracellular matrix on which the wing tissue depends to execute normal wing development. Planar tissue surface with tension and specific physicochemical factors of the extracellular matrix may be required to propagate morphogenic signals properly. These results can be explained by the assumption that chemical morphogens such as Wnt propagate on the dorsal side of the extracellular space of the hindwing. Alternatively, but not mutually exclusively, these results can be interpreted from the viewpoint of the distortion hypothesis and the induction model [3, 31–33]. The induction model that is integrated with the distortion hypothesis involves both mechanical signals (early stage) that follow a Newtonian equation to propagate [33] and chemical signals (late stage) that follow a short-range activation and a long-range inhibition, an essence of reaction-diffusion model [34–36]. The model proposes that the mechanical morphogenic signals are distortions of the planar epithelial sheet, which are translated into chemical signals (i.e., calcium waves and oscillations) that induce the expression of developmental regulatory genes such as Wnt [3].

2. Materials and methods

2.1. Butterfly samples and manipulations

The peacock pansy butterfly, *J. almana*, was used throughout this study. They were obtained from Ishigaki-jima Island, Okinawa, Japan. Eggs were collected from females, and larvae were reared at an ambient temperature using their natural host plants. No permissions were necessary to use these butterflies in biological research in Japan.

For all experimental procedures, the right forewing was lifted within 30 min after pupation, according to the previous studies that used this operation [8, 18, 21, 27–30]. After the operation of placement of either a ball or a covering material, the operated pupae were confined independently in a plastic container with a lid and placed at an ambient temperature until eclosion. After eclosion, the adult butterflies were frozen, and the wing color patterns were examined visually. The wing images were scanned using a Canon MG5730 scanner (Tokyo, Japan).

2.2. Ball placement

For the ball placement experiments, the forewing was first lifted and a stainless ball of 0.5 mm in diameter (Tsubaki Precision Balls, Tsubaki Nakashima, Katsuragi, Nara, Japan) was placed on the surface of the dorsal hindwing (**Figure 1A**). The forewing was then placed back to the original position. Thus, the ball was sandwiched between the forewing and the hindwing.

2.3. Contact treatments

For the contact experiments, a piece of transparent plastic film of polyvinylidene chloride (PVDC) for culinary use (Kurewrap, Kureha, Tokyo, Japan) was used to cover the wing surface with the operated wing upward (**Figure 1A, B**). The film was flexible enough to cover the entire surface of the exposed hindwing except the major eyespot. The anterior portion of the major eyespot was not exposed in this operation, and the posterior portion was exposed but might not been covered completely, because there was a small but disturbing physical gap between the epithelial surface and the surface of the pupal case that was not lifted. A different set of individuals were similarly covered with a piece of the plastic film, and the operated wing was placed downward (**Figure 1C**). Likewise, the dorsal hindwing surface was mounted on a Superfrost micro-glass slide (Matsunami Glass, Kishiwada, Osaka, Japan) (**Figure 1D**). This glass slide has a smooth surface that attaches to tissues, and it is thus frequently used for immunohistochemical analysis. In this case, the pupal body was lightly pushed onto a glass surface (**Figure 1E**). This way, the hindwing made a successful contact. Adult butterflies emerged from the operated pupae with severe forewing damage and color-pattern modifications of the operated dorsal hindwing (**Figure 1F**).

Additionally, a medical adhesive "white tape W129" with acrylic adhesives (Nichiban, Tokyo, Japan) was employed to cover the exposed hindwing surface. A piece of glassine paper coated with silicone resin (here called silicone-glassine paper) for culinary use (CGC Japan, Tokyo, Japan) was also used, on which the dorsal hindwing was placed (**Figure 1G, H**). In these treatments, the operated wing was placed downward. The adhesive tape and the silicone-glassine paper are not as flexible as the plastic film, and when a portion of the tissue was attached, the attached portion was confirmed from a horizontal view and from a non-attached side of the paper (**Figure 1G, H**).

2.4. Statistical analysis

Statistical analysis was not performed for the results of the major eyespot in comparison to the no-treatment group, because the characteristically disturbed eyespots by the ball were

http://dx.doi.org/10.5772/intechopen.70098

Figure 1. Mechanical disturbances applied to pupal hindwing in this study. (A) Operated pupa whose hindwing was covered with a piece of plastic film immediately after pupation. Asterisks indicate anterior (prospective eyespot) and posterior (prospective background) locations on which a 0.5-mm stainless ball was placed; in that case, the lifted forewing was placed back to the original position. (B) Adult hindwing color pattern seen through a piece of plastic film immediately before eclosion. Note that the posterior eyespot focus of the major eyespot is visible, but the anterior eyespot focus of the major eyespot is not visible because it is covered with the pupal case. The minor eyespot is found at the center of the exposed portion of the hindwing. Parafocal elements are also visible. (C) Operated pupa whose hindwing surface was covered with a piece of plastic film and placed down immediately after pupation. (D) Operated pupa whose hindwing surface was placed down onto a surface of a glass slide immediately after pupation. (E) Operated pupa as shown in D that was lightly pushed onto a glass surface. (F) An adult that eclosed from an operated pupa. Note that the hindwing major eyespot of the operated (right) side is smaller than that of the non-operated (left) side. (G) Operated pupa whose hindwing surface was placed down onto a piece of silicone-glassine paper. Only a central portion of the hindwing containing the minor eyespot and parafocal elements has a contact with the paper surface, and the prospective major eyespot has no physical contact. (H) The bottom image of a silicone-glassine paper that had an operated pupa. Only a central portion of the hindwing is attached to the paper. This attached portion contains the minor eyespot and parafocal elements, but not the major eyespot.

evident by their deformed shapes and because the covering operations were highly effective in almost all individuals treated (nearly 100%); high-level deformation and size reduction of the major eyespot were observed unilaterally. An exception was the film upward treatment, for which two-sided Fisher's exact test was performed in comparison to the film downward treatment and to the silicone-glassine paper treatment, using JSTAT 13.0 (2012) (Yokohama, Japan). There was no single case where such changes were obtained without an operation (no-treatment control here was $n = 76$ in this study alone, but such a case of changes has never been observed in many years of *J. almana* studies involving several hundred unoperated individuals). Similarly, statistical analysis for parafocal elements was not performed, because in the treatments using the adhesive tape and the silicone-glassine paper, changes of parafocal elements were seen in the majority of individuals (nearly 100%), whereas no such changes were observed in other treatments (0%). By contrast, Fisher's exact tests (two-sided results) were performed for the results of the minor eyespot in contrast to the data from no treatment, using JSTAT 13.0 (2012) (Yokohama, Japan).

3. Results

3.1. No-treatment control and forewing-lift control

The individuals without treatment (the no-treatment group) were first examined for their color-pattern symmetry or asymmetry of the major eyespot, the minor eyespot, and the parafocal elements between the right and the left hindwings in terms of their size and shape ($n = 76$). No extensive asymmetry was observed for the major eyespot (0%) and parafocal elements (0%). For the minor eyespot, however, 6 individuals out of 76 (8%) exhibited minor asymmetry. Similarly, as a basis of all experimental procedures that were performed in this study, a forewing-lift control experiment was performed, in which the forewing of a pupa was lifted and then placed back to the original position within a few minutes, and color patterns of these operated individuals were visually examined ($n = 20$). No change of color patterns was observed in the major eyespot (0%) and parafocal elements (0%) (**Figure 2A–C**). By contrast, 2 individuals out of 20 (10%) exhibited small-size asymmetry of the minor eyespot (**Figure 2A–C**). However, this asymmetry was not statistically significant in comparison to the non-treated individuals ($P = 1.0$).

3.2. Ball placement

A 0.5-mm ball was placed on the prospective major eyespot of the dorsal hindwing (**Figure 1A**). Because the exposure was limited to the posterior side of the major eyespot (**Figure 1B**), the ball was most likely placed on the posterior side of the major eyespot ($n = 21$). Ten treated individuals out of 21 (48%) exhibited irregular changes of the major eyespot (**Figure 2D–F**).

Likewise, a 0.5-mm ball was placed in the central background position of the dorsal hindwing (**Figure 1A**). The ball had no physical contact with the major eyespot ($n = 21$). Three treated individuals out of 21 (14%) exhibited irregular minor changes of the major eyespot in its anterior side, a remote place from the ball (**Figure 2G–I**).

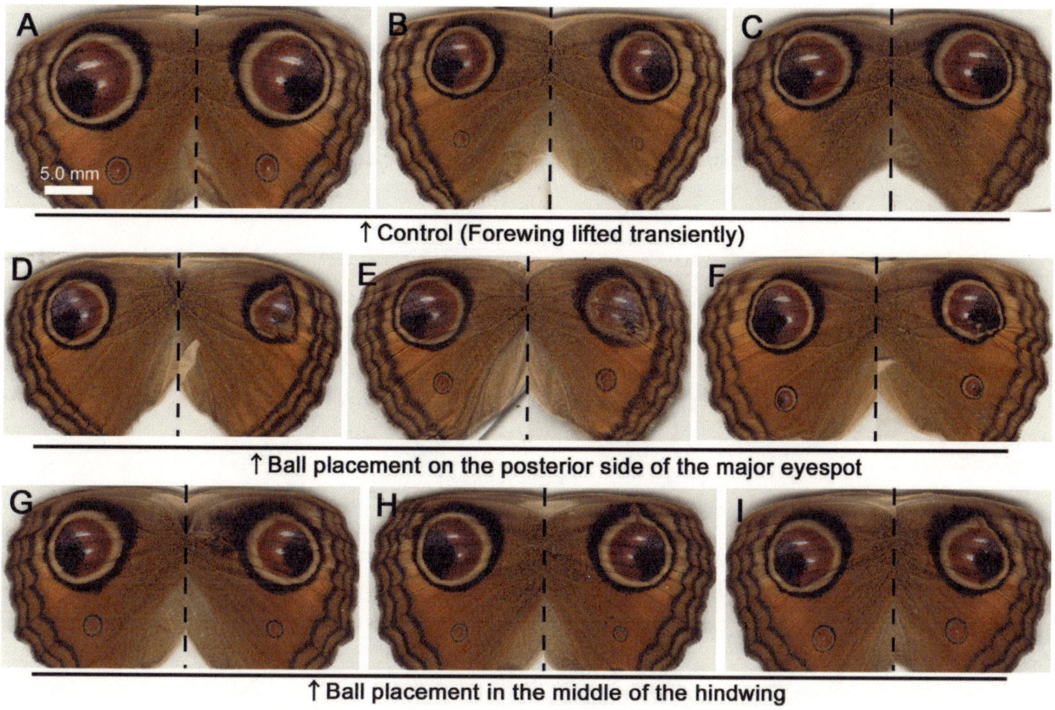

Figure 2. Forewing-lift control and ball placement. The right hindwing was operated and the left hindwing of the same individual was shown in every panel as the internal control for color-pattern comparison. (A–C) Control individuals. The forewing was lifted transiently and placed back to the original position. No changes of the color patterns are observed except that in B the right minor eyespot is smaller than the left one. However, this is not statistically significant in comparison to the no-treatment group ($P = 1.0$). (D–F) Ball placement on the posterior side of the major eyespot. Extensive deformation of the major eyespots is observed. (G–I) Ball placement in the middle of the hindwing. Color-pattern modifications are observed in the proximal (G) or anterior (H, I) part of the major eyespot.

3.3. Plastic film over the hindwing

After the forewing-lift procedure, the exposed hindwing was covered with a piece of transparent plastic film ($n = 27$) (**Figure 1A, B**). The operated right side was placed upward so that pressure on the hindwing surface was minimal. High-level size reduction with deformation of the major eyespot was observed in 20 treated individuals out of 27 (74%) (**Figure 3A–C**). Low-level size reduction or deformation was observed in 4 out of 27 (15%) (**Figure 3D**). Together, 24 out of 27 (89%) showed a change of the major eyespot. Even in the cases of the high-level reduction, the white spots inside the major eyespot were not affected much, and parafocal elements did not change, either. In the case of the minor eyespot, 3 individuals out of 20 (15%) that were visually judged clearly showed size reduction (1 individual; **Figure 2A**) and size enlargement (2 individuals; **Figure 2B, C**). However, these changes of the minor eyespot were not statistically significant in comparison to the no-treatment group ($P = 0.39$).

To examine if a light pressure on the hindwing due to its own weight may change color patterns, the exposed hindwing that was covered with a piece of plastic film was placed downward on a solid surface ($n = 14$) (**Figure 1D, E**). All 13 individuals out of 14 (93%) showed high-level reduction of the major eyespot (**Figure 3E, F**). As an exception, one individual did not show any change. The downward configuration did not show significant difference from the upward configuration when the high-level and low-level changes were not treated as distinguished categories ($P = 1.0$). Even when only the high-level changes were compared, the downward configuration did not show significant difference from the upward configuration ($P = 0.23$).

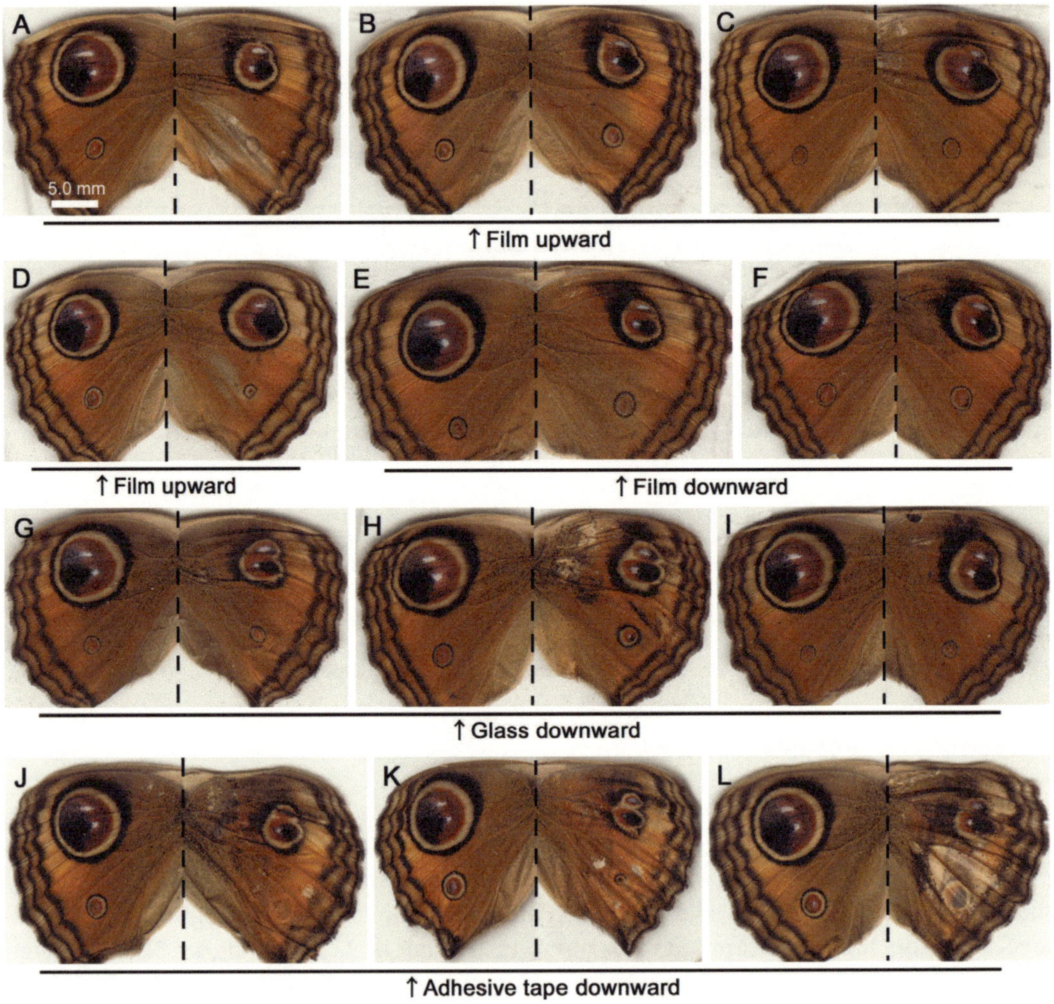

Figure 3. Film, glass, and adhesive tape experiments. The right hindwing was operated and the left hindwing of the same individual was shown in every panel as the internal control for color-pattern comparison. In all cases shown, the right major eyespot was deformed and reduced in size. (A–D) The hindwing surface was covered with a piece of plastic film and the treated wing was placed upward. The right minor eyespot was either reduced (A, D) or enlarged (B, C) in size. (E, F) The hindwing surface was covered with a piece of plastic film and the treated wing was placed downward. The right minor eyespot was enlarged in these individuals. (G–I) The hindwing surface was mounted on a piece of glass slide. The minor eyespot was reduced in size (G), showed the white spot inside (H), or showed no change (I). (J–L) The hindwing surface was mounted on a piece of adhesive tape. The minor eyespot was reduced in size in (K), but the scales of the minor eyespot in (J) and (L) were removed, which made observations of the minor eyespot impossible in these individuals. In (L), the scale removal is extensive; the ventral side is seen through.

Again, parafocal elements did not change, but 3 individuals (2 enlargements, **Figure 3E**, **F**; and 1 reduction, **Figure 3D**) out of 14 (21%) showed changes in the minor eyespot. However, these changes of the minor eyespot were not statistically significant in comparison to the no-treatment group ($P = 0.14$).

3.4. Hindwing placement on a glass slide

To examine the possibility that the covering materials may affect color patterns, the exposed hindwing was directly placed on the surface of a glass slide (**Figure 1D**). The hindwing was lightly pushed on the glass surface so that the hindwing could make a direct contact with a glass surface at least at that time point (**Figure 1E**). Thus, the operated side was placed downward ($n = 15$). Because the glass surface is rigid in contrast to the flexible film, the major eyespot may not have maintained a contact with a glass surface, as with the case of the white adhesive tape and the silicone-glassine paper (see subsequent experiments).

After the glass treatment, high-level changes with deformation of the major eyespot were observed in all 15 treated individuals (100%) (**Figure 3G–I**). Although not quantitative, the level of size reduction was also likely more severe than the previous film treatments. No change was observed in parafocal elements. The minor eyespot was affected in 5 out of 14 (36%). Among them, 3 showed reduction (**Figure 3G**) and the other 2 showed white spot emergence (**Figure 3H**). The minor eyespot changes were statistically significant in comparison to the no-treatment group ($P = 0.012$). Separately, the size reduction ($P = 0.0031$) and the white spot appearance ($P = 0.023$) were both significant. The surface rigidity or physicochemical nature of the glass slide might have contributed to these color-pattern changes of the minor eyespot.

3.5. Hindwing placement on a piece of adhesive tape

Here, it was hypothesized that surface adhesion may contribute to color-pattern changes. A piece of adhesive tape was used to cover the surface of the exposed hindwing. However, in this treatment, it was confirmed that there was no direct contact with the major eyespot. That is, the major eyespot was not physically covered with the tape. By contrast, the minor eyespot was completely covered. This configuration was the same as the silicone-glassine paper treatment (**Figure 1G**, **H**). Thus, the effects on the major eyespot are basically from no-covering material. But the effects on the minor eyespot are from a covering material on it.

In all 14 individuals that eclosed (including 3 individuals that formed complete adult wings in pupae but failed to exit from the pupal case), high-level reduction of the major eyespot was observed (100%) (**Figure 3J–L**). Although not quantitative, the level of reduction appeared to be more severe than the previous treatments. Interestingly, the minor eyespots were also reduced in all of these individuals ($n = 13$) (100%), although one individual cannot be judged because of the removal of scales of the minor eyespot upon eclosion. This result was statistically significant ($P < 0.0001$). Because of high adhesiveness of the tape, scales were often removed at the time of eclosion around the minor eyespot (**Figure 3J–L**). The scale removal demonstrated the direct (or nearly direct) adhesion of the tape to scales. Parafocal elements and submarginal bands were thickened and somewhat displaced proximally in all individuals (100%), which is reminiscent of the temperature-shock-type (TS-type) modifications [24, 38–40].

3.6. Hindwing placement on a sheet of silicone-glassine paper

To gain further insights into mechanical and physicochemical factors for color-pattern determination, the exposed hindwing surface was placed on a sheet of silicone-glassine paper ($n = 26$) (**Figure 1G, H**). The silicone-glassine paper was used here because it is not supposed to stick to the wing surface tightly, in contrast to the adhesive tape used earlier. Contrary to the expectation, results were similar to those of the adhesive tape. In all 26 individuals, high-level reduction of the major eyespot was observed (100%) (**Figure 4A–I**).

Interestingly, the minor eyespot changes in coloration and size were observed in 21 individuals out of 24 (2 individuals were not possible to judge because of breakage of the wings during eclosion and manipulation) in the silicone-glassine paper treatment (**Figure 4A–I**). This result was statistically significant ($P < 0.0001$). Similarly, parafocal elements changed in size and location, reminiscent of the TS-type modifications [24, 38–40], in 14 treated individuals out of 20 (6 individuals were not possible to judge because of breakage of the wings during eclosion and manipulation) (**Figure 4A–I**). In addition, small ectopic spots or ring structures were often observed in the background.

3.7. Response profiles of color-pattern elements

On the basis of the experimental results on the number of individuals that exhibited color-pattern changes, response profiles of the major eyespot, the minor eyespot, and parafocal

Figure 4. Silicone-glassine paper experiment. The right hindwing was operated and the left hindwing of the same individual was shown in every panel as the internal control for color-pattern comparison. In all cases, not only the major eyespot but also parafocal elements and the minor eyespot were affected. The minor eyespot modifications are different from that of the major eyespot. But overall, the modifications are reminiscent of those induced by tungstate injection or temperature-shock treatment [24, 38–40].

elements were obtained (**Figure 5A**). The major eyespot was always disrupted by any treatments; this is probably because no mode of treatment covered the prospective major eyespot area, with a possible exception of the film treatment. Indeed, when the high-level changes of the film upward treatment were compared to the silicone-glassine paper treatment, their difference was statistically significant ($P = 0.010$). This result likely means that the upward film treatment physically covered the exposed posterior portion of the major eyespot at least in some individuals and that this film cover functionally mimicked the natural extracellular matrix for the hindwing tissue to some extent.

In contrast to the major eyespot, the minor eyespot and parafocal elements were firmly covered by the covering materials, which mean that the effects on the minor eyespot and parafocal element may be caused by physicochemical properties of the materials. Parafocal elements were affected only by the adhesive tape and the silicone-glassine paper.

The response profiles of the minor eyespot were further obtained in terms of three types of color-pattern changes: size reduction, size enlargement, and appearance of the white spot at the center (**Figure 5B**). Among them, reduction was the most frequent change in the glass ($P = 0.0031$), the adhesive tape ($P < 0.0001$), and the silicone-glassine paper ($P < 0.0001$).

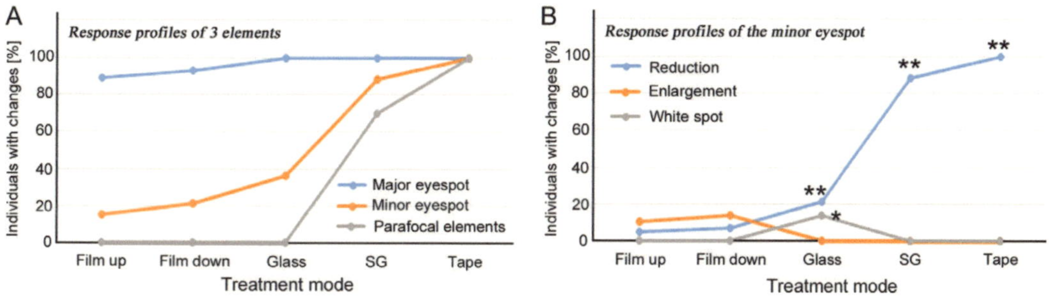

Figure 5. Response profiles of color-pattern elements. "SG" and "tape" indicate the silicone-glassine paper treatment and the adhesive tape treatment, respectively. (A) Profiles of three elements. For the film upward treatment, both high-level and low-level changes were included without distinction. The major eyespot is highly responsive to all the treatment modes. Parafocal elements are sensitive to the two modes, the adhesive tape and the silicone-glassine paper. These are highly significant results without doubt in comparison to the no-treatment group. (B) Profiles of the minor eyespot. Three different response patterns are recognized, and they are profiled in response to the treatment modes. Size reduction is prominent in the adhesive tape (**$P < 0.0001$) and the silicone-glassine paper (**$P < 0.0001$), but it can also be seen in the glass treatment (**$P = 0.0031$). These results are statistically significant. The white spot appearance in the glass treatment is also significant (*$P = 0.023$). By contrast, size reduction and enlargement seen in the film upward treatment ($P = 0.39$) and the film downward treatment ($P = 0.14$) are not statistically significant.

4. Discussion

4.1. Overview of this study

In this study, different types of mechanical distortions and adhesions on developing pupal hindwing tissues were introduced. The present study is composed of three parts that require independent interpretations: (1) the response of the major eyespot to the ball placement, (2) the response of the major eyespot to no-covering material via the forewing-lift method, and (3) the response of the minor eyespot (together with the parafocal elements) to various covering materials. Collectively, however, the present results demonstrated that artificially introduced mechanical distortions and properties of contact surface affect the final color patterns in butterfly wings.

4.2. Ball placement and physical damage

The degrees of size reduction in the major eyespot in response to the ball treatment may be compared with the damage-induced changes in the previous study [26]. When the anterior eyespot focus was physically damaged by a stainless needle, the major eyespot was reduced in size not only in the anterior side but also in the posterior side, suggesting synergistic interactions of signals from two adjacent organizers. When the posterior eyespot focus was damaged, similar effect was observed, but it was much less effective [26]. In the present study, the ball placed on the posterior portion of the major eyespot appears to be at least as effective as physical damage at the posterior focus, suggesting the importance of distortion in developmental fate determination. Assuming that the ball placement did not kill epithelial cells, the present results suggest that necrotic cell death caused by physical damage is not necessary to induce color-pattern changes. The ball placement on the background was less influential, but interestingly, it induced irregular local extrusion of the major eyespot, suggesting that the mechanical distortion may impose a long-range effect on the major eyespot. On the other hand, a small degree of wing-wide pressure on the hindwing in the downward configuration with the plastic film coverage did not change color patterns at the anterior side, suggesting that a local distortion of the planar tissue may be more important than a wing-wide pressure (i.e., distortion) to cause changes in color patterns.

4.3. Extracellular environment of the dorsal hindwing surface

It is important to understand the extracellular environment of the hindwing tissue before discussing possible interpretations of the experimental results of various covering materials. The hindwing dorsal surface, when the forewing was lifted immediately after pupation, may not be covered with cuticles. If any, that cuticle coverage may be very thin. Alternatively, the forewing-lift operation and/or coverage with artificial materials may completely inhibit or reduce the cuticle formation process on the surface of the hindwing. To be consistent with this idea, a long-term hindwing exposure without any coverage after the operation makes them die from being dried [21]. This was also confirmed in the present study; all the operated pupae (n = 24) with an exposed hindwing without any coverage (but the ventral forewing was covered with a piece of plastic film) died without development of color patterns (100%). Furthermore, the adhesive tape treatment removed many scales from the dorsal hindwing upon eclosion, probably because the adhesive tape was sticky enough to bind scales directly and strongly. Thus, it is likely that the extracellular side of the hindwing tissue cells was directly exposed to covering materials.

4.4. Response of the major eyespot

The major eyespot of the dorsal hindwing was sensitive to the operations performed in this study. Use of various covering materials with different rigidity, adhesiveness, surface smoothness, and chemical composition resulted in miniaturization of the major eyespot. But in the adhesive tape and the silicone-glassine paper treatments (and probably also in the plastic film and glass treatments), the posterior side of the major eyespot was not in contact with anything. Because of curvature of the hindwing tissue and a physical gap between the surface of the hindwing tissue and the pupal case of the most ventral part, even the flexible plastic film cannot completely make a contact with the major eyespot. This configuration was clearly confirmed in the adhesive white tape and the silicone-glassine paper treatments. Furthermore, it is to be noted that the major eyespot in this butterfly could not be completely exposed by the forewing-lift method; the anterior portion was always under the pupal case. These facts likely explain that the results of various covering treatments were virtually identical for the major eyespot.

It is surprising that the miniaturization of the major eyespot by the present operations is more efficient than the physical damage treatment [26], despite the fact that the present operations are less invasive. Likely interpretations would be that the major eyespot organizers need extracellular supporting materials to propagate morphogenic signals and that the anterior side alone that was covered with the pupal case cannot expand without the help of the posterior side. These interpretations are consistent with the previous chapter that describes synergistic signal amplification and expansion processes in this eyespot [26]. Indeed, the anterior side of the major eyespot appeared to be more sensitive to the present treatments and also to physical damage [26] than the posterior side despite the fact that the anterior side is physically hidden. It is to be noted that the upward film treatment was the least effective to induce changes. And there is a possibility that this treatment covered the posterior part of the major eyespot at least in some individuals because of its flexibility. Therefore, for the morphogenic signals to propagate efficiently, a covering material is required. However, judging from the effects of various covering materials on the minor eyespot, the covering materials should have certain physicochemical properties to support normal propagation of morphogenic signals. In this sense, the plastic film that did not affect the minor eyespot significantly is ideal and may be similar to the normal extracellular matrix of the hindwing epithelium in *J. almana*.

As a general tendency, the proximal side of the major eyespot showed a fusion of the signals from the anterior and posterior organizers, whereas the distal side often showed a separation of the two. Signals may be more expandable to the proximal side. In the reduced major eyespot, the size of the white spots was not affected much in the film and glass treatments, suggesting an uncoupling behavior of the white spots from the rest of the eyespot. Similar uncoupling behavior of white spots has been shown in *Calisto* butterflies [37]. Similar to the white spots, the minor eyespot and parafocal elements were not affected very much by the film treatment. However, the glass treatment significantly induced the size reduction and the white spot induction of the minor eyespot. This induction may be specific to the glass surface physical chemistry, but the low level of pressure applied in this particular treatment may be a reason.

4.5. Adhesive tape and silicone-glassine paper treatments

In contrast to the major eyespot, the minor eyespot was in direct contact with the covering materials. Also in contrast to the film and glass treatments, which did not induce significant changes in the minor eyespot, both the adhesive tape and the silicone-glassine paper treatments unexpectedly induced extensive modifications of the minor eyespot and parafocal elements, in addition to the reduction of the major eyespot. In these two treatments, the size reduction of the minor eyespot was statistically significant, suggesting the importance of a functional contact surface in expanding morphogenic signals for eyespots. Furthermore, size reduction of the white spot inside the major eyespot was prominent in the two treatments. If chemical morphogens are secreted to the apical extracellular side of epithelial cells, chemical morphogen transport would be disrupted by different covering materials. The present results using various covering materials do not contradict with this idea.

4.6. Similarity to the TS-type modifications

The overall phenotype, the displaced and diffused parafocal elements and the smaller major and minor eyespots induced by the adhesive tape and the silicone-glassine paper, is similar to the tungstate-injected phenotype, or more generally temperature-shock-type (TS-type) modifications that were demonstrated in this and other nymphalid butterfly species [22–24, 38–40]. The tungstate treatment and temperature-shock treatment have been known to induce characteristic wing-wide color-pattern modifications in this species [24]; eyespots became smaller and parafocal elements are diffused and dislocated proximally toward the eyespot focus. It appears that the adhesive tape and the silicone-glassine paper treatments were as effective as the injection of tungstate to produce the TS-type modifications or their similar ones in this species. This fact suggests that the mechanisms for the size reduction by covering materials and by tungstate injection may basically be similar. In that case, tungstate, cold shock, and the cold-shock hormone may act on the extracellular matrix of the wing tissue.

However, there is an important difference between the contact treatments and the tungstate and its related treatments. In the contact treatments, the minor eyespot appeared to be more sensitive than parafocal elements (note that a comparison to the major eyespot is irrelevant, because it was not in contact with anything). Parafocal elements were not modified in the glass treatment but the minor eyespot was. Morphogenic signals for parafocal elements had already been released by the time of the treatments, but signals for the minor eyespot had not [32]. It appears that the contact treatments affect the early phase of signals than the moving phase. Tungstate and its related treatments affect in an opposite way. In this sense, these two modes of treatments are different. The reason for this difference is unknown. A possible speculation is as follows. During development, the wing tissue shows a slow contraction cycles [28], which may contribute to an adjustment of the physical properties (including that of the extracellular matrix) of developing epithelial tissues. Because the epithelial tissue is covered by inflexible materials in the covering experiments (except for the film treatment), this contraction movement may be inhibited, affecting morphogenic signals to be released and propagate. Morphogenic signals that were released already may not be affected much, because it is less dependent on the contractive movement anymore.

Interestingly, heparin, chondroitin sulfates, and dextran sulfate that could act extracellularly are also able to induce TS-type modifications [41]. Because heparin sulfate proteoglycans play an important role in Wnt signaling [42–49], because Wnt family proteins are thought to be chemical morphogens for butterfly wing color patterns [16], and because TS-type modifications may be attained by molecular changes of the extracellular matrix, various treatments that induce TS-type modifications may cause the reduction of the extracellular movement of Wnt family proteins. On the other hand, Wnt may be transmitted via membranous structures such as cytonemes [50–54] and argosomes [44] through intercellular spaces that are filled with hemolymph inside the tissue. Cytoneme-like structures were reported in the developing butterfly wing tissue [14, 29].

The distortion hypothesis and induction model (see subsequently) posit mechanical morphogens, but they do not deny (but incorporated) chemical morphogens such as Wnt family proteins that would play an important role in finalizing the adult color patterns. Mechanical signals may be released first from organizers, but they should readily be translated into chemical signals that act locally. Activation of TGF-β in the extracellular matrix is executed by mechanical forces medicated by integrins and other extracellular matrix molecules [55]. Interestingly, TGF-β has been considered a candidate morphogen in butterfly wings [15].

4.7. The induction model and the distortion hypothesis

The induction model has been proposed to explain processes of color-pattern determination in butterfly wings, based on several lines of evidence including color-pattern comparisons among many butterfly species [31, 32], experimentally induced color-pattern changes [25], scale-size distribution patterns [21, 27], morphological and histochemical analyses of pupal wings [12], mathematical modeling [33], and developmental real-time imaging [14, 28, 29]. In this model, morphogenic signals are released as slow decelerating wave pulses from organizers, and the locations of their settlement then act as the secondary organizers [3]. However, identity of the wave signals has been enigmatic. The present study has suggested that one possible candidate is mechanical distortions of the epithelial tissues and highlighted the importance of the extracellular matrix as a medium for mechanical or chemical signals.

The distortion hypothesis has been proposed, in which the putative wave signals were explained as mechanistic distortions of the wing epithelial tissues [3]. Cuticle spots are likely sources of distortions, and distortions slowly propagate radially with decelerating motion. Distorted immature scale cells are activated by calcium waves through a stretch-sensitive calcium channel. Distortions act as a ploidy signal, and the degrees of polyploidy of the epithelial cells determine the final coloration of a given scale [27]. This distortion hypothesis can explain the nature of morphogenic signals that have been proposed in the induction model of positional information in butterfly wings. To generate and propagate the wave signals, the planar epithelial sheet and its supporting materials (i.e., the extracellular matrix) with their appropriate physicochemical properties may be required.

Surface rigidity that is conferred by the extracellular matrix may play an important role in development in general by giving mechanical supports for cells [1]. In *Drosophila,* the apical surface of wing epithelial cells changes its morphology, and this morphology acts as a template to produce a rigid dorsal cuticle. After that, a flexible ventral cuticle is produced, which is then molded on the inner surface of the rigid dorsal cuticle [56, 57]. A similar mechanism has been proposed in the development of butterfly scales [58].

5. Conclusions

The present study provided experimental evidence that mechanical force and physicochemical properties of extracellular matrix contribute to morphogenic signal propagation, focusing on the hindwing color patterns of the peacock pansy butterfly. These results point to the importance of an appropriate tension and the extracellular milieu that the planar wing epithelium has. Mechanical distortions and physicochemical properties of the extracellular matrix may be functional mediators of long-range morphogenic signals in butterfly wings.

Acknowledgements

The author thanks members of the BCPH Unit of Molecular Physiology at the University of the Ryukyus for discussions. This work was supported by the basic research fund from University of the Ryukyus and by JSPS KAKENHI, Grant-in-Aid for Scientific Research (C), Grant Number 16K07425. The author declares that he has no competing interests.

Author details

Joji M. Otaki

Address all correspondence to: otaki@sci.u-ryukyu.ac.jp

The BCPH Unit of Molecular Physiology, Department of Chemistry, Biology and Marine Science, Faculty of Science, University of the Ryukyus, Senbaru, Nishihara, Okinawa, Japan

References

[1] Iskratsch T, Wolfenson H, Sheetz M. Appreciating force and shape—The rise of mechanotransduction in cell biology. Nature Reviews Molecular Cell Biology. 2014;**15**:825-833

[2] Gilmour D, Rembold M, Leptin M. From morphogen to morphogenesis and back. Nature. 2017;**541**:311-320

[3] Otaki JM. Self-similarity, distortion waves, and the essence of morphogenesis: A generalized view of color pattern formation in butterfly wings. In: Sekimura T, Nijhout HF, editors. Integrative Approach to the Analysis of the Diversity and Evolution of Butterfly Wing Patterns. Singapore: Springer Nature Singapore; 2017. pp. 119-152

[4] Wolpert L. Positional information and the spatial pattern of cellular differentiation. Journal of Theoretical Biology. 1969;**25**:1-47

[5] Wolpert L. One hundred years of positional information. Trends in Genetics. 1996;**12**: 359-364

[6] Nijhout HF. The Development and Evolution of Butterfly Wing Patterns. Washington, DC: Smithsonian Institution Press; 1991. p. 297

[7] Nijhout HF. Elements of butterfly wing patterns. Journal of Experimental Zoology. 2001;**291**:213-225

[8] Dhungel B, Otaki JM. Local pharmacological effects of tungstate on the color-pattern determination of butterfly wings: A possible relationship between the eyespot and parafocal element. Zoological Science. 2009;**26**:758-764

[9] Otaki JM. Color-pattern analysis of parafocal elements in butterfly wings. Entomological Science. 2009;**12**:74-83

[10] Otaki JM. Colour pattern analysis of nymphalid butterfly wings: Revision of the nymphalid groundplan. Zoological Science. 2012;**29**:568-576

[11] Taira W, Kinjo S, Otaki JM. The marginal band system in the nymphalid butterfly wings. Zoological Science. 2015;**32**:38-46

[12] Otaki JM, Ogasawara T, Yamamoto H. Morphological comparison of pupal wing cuticle patterns in butterflies. Zoological Science. 2005;**22**:21-34

[13] Taira W, Otaki JM. Butterfly wings are three-dimensional: Pupal cuticle focal spots and their associated structures in *Junonia* butterflies. PLoS One. 2016;**11**:e0146348

[14] Iwasaki M, Ohno Y, Otaki JM. Butterfly eyespot organiser: *In vivo* imaging of the prospective focal cells in pupal wing tissues. Scientific Reports. 2017;**7**:40705

[15] Monteiro A, Glaser G, Stockslager S, Glansdrop N, Ramos D. Comparative insights into questions of lepidopteran wing pattern homology. BMC Developmental Biology. 2006;**6**:52

[16] Martin A, Reed RD. *Wnt* signaling underlies evolution and development of the butterfly wing pattern symmetry systems. Developmental Biology. 2014;**395**:367-378

[17] Monteiro A, Chen B, Ramos DM, Oliver JC, Tong X, Guo M, Wang WK, Fazzino L, Kamal F. *Distal-less* regulates eyespot patterns and melanization in *Bicyclus* butterflies. Journal of Experimental Zoology. Part B: Molecular and Developmental Evolution. 2013;**320**:321-331

[18] Adhikari K, Otaki JM. A single-wing removal method to assess correspondence between gene expression and phenotype in butterflies: The case of *Distal-less*. Zoological Science. 2016;**33**:13-20

[19] Dhungel B, Ohno Y, Matayoshi R, Iwasaki M, Taira W, Adhikari K, Gurung R, Otaki JM. Distal-less induces elemental color patterns in *Junonia* butterfly wings. Zoological Letters. 2016;**2**:4

[20] Zhang L, Reed RD. Genome editing in butterflies reveals that *spalt* promotes and *Distal-less* represses eyespot colour patterns. Nature Communications. 2016;**7**:11769

[21] Kusaba K, Otaki JM. Positional dependence of scale size and shape in butterfly wings: Wing-wide phenotypic coordination of color-pattern elements and background. Journal of Insect Physiology. 2009;**55**:174-182

[22] Mahdi SH, Gima S, Tomita Y, Yamasaki H, Otaki JM. Physiological characterization of the cold-shock-induced humoral factor for wing color-pattern changes in butterflies. Journal of Insect Physiology. 2010;**56**:1022-1031

[23] Mahdi SHA, Yamasaki H, Otaki JM. Heat-shock-induced color-pattern changes of the blue pansy butterfly *Junonia orithya*: Physiological and evolutionary implications. Journal of Thermal Biology. 2011;**36**:312-321

[24] Otaki JM. Reversed type of color-pattern modifications of butterfly wings: A physiological mechanism of wing-wide color-pattern determination. Journal of Insect Physiology. 2007;**53**:526-537

[25] Otaki JM. Artificially induced changes of butterfly wing colour patterns: Dynamic signal interactions in eyespot development. Scientific Reports. 2011;**1**:111

[26] Iwasaki M, Otaki JM. Synergistic damage response of the double-focus eyespot in the hindwing of the peacock pansy butterfly. In: Perveen FK, editor. Lepidoptera. Rijeka, Croatia: InTech; 2017

[27] Iwata M, Otaki JM. Spatial patterns of correlated scale size and scale color in relation to color pattern elements in butterfly wings. Journal of Insect Physiology. 2016;**85**:32-45

[28] Iwata M, Ohno Y, Otaki JM. Real-time *in vivo* imaging of butterfly wing development: Revealing the cellular dynamics of the pupal wing tissue. PLoS One. 2014;**9**:e89500

[29] Ohno Y, Otaki JM. Live cell imaging of butterfly pupal and larval wings *in vivo*. PLoS One. 2015;**10**:e0128332

[30] Ohno Y, Otaki JM. Spontaneous long-range calcium waves in developing butterfly wings. BMC Developmental Biology. 2015;**15**:17

[31] Otaki JM. Color-pattern analysis of eyespots in butterfly wings: A critical examination of morphogen gradient models. Zoological Science. 2011;**28**:403-413

[32] Otaki JM. Generation of butterfly wing eyespot patterns: A model for morphological determination of eyespot and parafocal element. Zoological Science. 2011;**28**:817-827

[33] Otaki JM. Structural analysis of eyespots: Dynamics of morphogenic signals that govern elemental positions in butterfly wings. BMC Systems Biology. 2012;**6**:17

[34] Gierer A, Meinhardt H. A theory of biological pattern formation. Kybernetik. 1972;**12**:30-39

[35] Meinhardt H, Gierer A. Pattern formation by local self-activation and lateral inhibition. BioEssays. 2000;**22**:753-760

[36] Kondo S, Miura T. Reaction-diffusion model as a framework for understanding biological pattern formation. Science. 2010;**329**:1616-1620

[37] Iwata M, Otaki JM. Focusing on butterfly eyespot focus: Uncoupling of white spots from eyespot bodies in nymphalid butterflies. SpringerPlus. 2016;**5**:1287

[38] Otaki JM. Color-pattern modifications of butterfly wings induced by transfusion and oxyanions. Journal of Insect Physiology. 1998;**44**:1181-1190

[39] Otaki JM. Physiologically induced color-pattern changes in butterfly wings: Mechanistic and evolutionary implications. Journal of Insect Physiology. 2008;**54**:1099-1112

[40] Otaki JM, Ogasawara T, Yamamoto H. Tungstate-induced color-pattern modifications of butterfly wings are independent of stress response and ecdysteroid effect. Zoological Science. 2005;**22**:635-644

[41] Serfas MS, Carroll SB. Pharmacologic approaches to butterfly wing patterning: Sulfated polysaccharides mimic and antagonize cold shock and alter the interpretation of gradients of positional information. Developmental Biology. 2005;**287**:416-424

[42] Erickson JL. Formation and maintenance of morphogen gradients: An essential role for the endomembrane system in *Drosophila melanogaster* wing development. Fly. 2011;**5**: 266-271

[43] Hufnagel L, Kreuger J, Cohen SM, Shraiman BI. On the role of glypicans in the process of morphogen gradient formation. Developmental Biology. 2006;**300**:512-522

[44] Greco V, Hannus M, Eaton S. Argosomes: A potential vehicle for the spread of morphogens through epithelia. Cell. 2001;**106**:633-645

[45] Fuerer C, Habib SJ, Nusse R. A study on the interactions between heparan sulfate proteoglycans and Wnt proteins. Developmental Dynamics. 2010;**239**:184-190

[46] Binari RC, Staveley BE, Johnson WA, Godavarti R, Sasisekhran R, Manoukian AS. Genetic evidence that heparin-like glycosaminoglycans are involved in wingless signaling. Development. 1997;**124**:2623-2632

[47] Baeg GH, Lin X, Khare N, Baumgartner S, Perrimon N. Heparan sulfate proteoglycans are critical for the organization of the extracellular distribution of wingless. Development. 2001;**128**:87-94

[48] Lin X, Perrimon N. Developmental roles of heparin sulfate proteoglycans in *Drosophila*. Glycoconjugate Journal. 2002;**19**:363-368

[49] Lin X. Functions of heparan sulfate proteoglycans in cell signaling during development. Development. 2004;**131**:6009-6021

[50] Gradilla AC, Guerrero I. Cytoneme-mediated cell-to-cell signaling during development. Cell and Tissue Research. 2013;**352**:59-66

[51] Kornberg TB, Roy S. Communicating by touch—Neurons are not alone. Trends in Cell Biology. 2014;**24**:370-376

[52] Kornberg TB, Roy S. Cytonemes as specialized signaling filopodia. Development. 2014;**141**: 729-736

[53] Roy S, Kornberg TB. Paracrine signaling mediated at cell-cell contacts. BioEssays. 2015;**37**:25-33

[54] Stanganello E, Scholpp S. Role of cytonemes in Wnt transport. Journal of Cell Science. 2016;**129**:665-672

[55] Robertson IB, Rifkin DB. Unchaining the beast; insights from structural and evolutionary studies on TGFβ secretion, sequestration, and activation. Cytokine & Growth Factor Reviews. 2013;**24**:355-372

[56] Valentine M, Collier S. Planar cell polarity and tissue design. Shaping the *Drosophila* wing membrane. Fly. 2011;**5**:316-321

[57] Belalcazar AD, Doyle K, Hogan J, Neff D, Collier S. Insect wing membrane topology is determined by the dorsal wing epithelium. G3. 2013;**3**:5-8

[58] Ghiradella H. Structure of butterfly scales: Patterning in an insect cuticle. Microscopy Research and Technique. 1994;**27**:429-438

Chapter 6

Mitochondrial Genomes of Lepidopteran Insects Considered Crop Pests

Viviana Ramírez-Ríos, Javier Correa Alvarez and Diego Villanueva-Mejia

Additional information is available at the end of the chapter

http://dx.doi.org/10.5772/intechopen.71158

Abstract

In this chapter, the complete mitochondrial genome of Guatemalan potato moth, *Tecia solanivora* (Povolny, 1973) (Lepidoptera: Gelechiidae) is presented as a model to understand how to characterize and study a mitogenome in insects. It was sequenced, analyzed, and compared with other lepidopteran insects. *T. solanivora* mitogenome is a circular double-stranded molecule, typically found in insects and containing 37 genes, all them well described over the other lepidopteran mitogenomes sequenced. Interestingly, in this mitogenome was found a gene arrangement in the tRNA-Met gene different from the ancestral arrangement, but commonly present in insect mitogenomes. Other important characteristics are the high A + T-biased and negative AT- and GC-skews contents, but also unusual canonical start codons in 12 protein-coding genes and an incomplete stop codon in the cytochrome oxidase subunit II gene consisting of just a Thymine. Another common feature shared with lepidopteran mitogenomes is the A + T-rich region. It is characterized by having 325 bb, the 'ATAGA' motif, a 17 bp poly (T) stretch and a $(AT)_8$ element preceded by the 'ATTTA' motif. Likewise, this mitogenome has 21 intergenic spacer regions. In addition, an update about other recent mitogenomes research done mainly over lepidopteran insects considered crop pests is presented. On the other hand, a novel development based on induced mutations by CRISPR-Cas9 in the mitogenomes seeking applicable capability for pest control is shown. The utility of this study is to improve scientific databases and support future studies of population genetic in lepidopteran.

Keywords: mitogenomes, mitochondrial genome, crop pests, lepidopteran, insects

1. Introduction

Crop loss is a function of one or more biotic factors, each of which may be contributing to a reduction in yield, whereas yield loss is the reduction in yield caused by a single pathogen or a pest [1]. Even so, there is no doubt that crop losses due to pests and diseases are a major threat to incomes of rural families and to food security worldwide [2]. Although there are a large number of different living organisms that affect agricultural crops (biotic factors) and therefore they are called pests or pathogens, organisms from Lepidoptera order within Insecta class are considered one of the most economically important pests due to huge crop losses caused by them, crop losses, in terms of quantity and quality that can occur in the field (pre-harvest) or in the storage (post-harvest) [3].

Lepidoptera (moths and butterflies) is the second largest order in Insecta, is species-rich containing over 155,000 described species, and occurs in nearly all regions and a wide variety of habitats [4]. A combination of features has conspired to render the Lepidoptera one of the most studied groups of organisms; on account of this, research on lepidopteran insects has been carried out during the past century. Nevertheless, only few years ago, scientists are seeking answers on genomes as a key to revalidate previously generated data or redirect mechanisms of pest control. In this context, mitochondrial genomes (mtgenomes or mitogenomes) are very important subject for different scientific disciplines including, among others, animal health, comparative and evolutionary genomics, molecular evolution, phylogenetic, population genetics, and biogeographic studies [5, 6]. Therefore, it is not surprising that

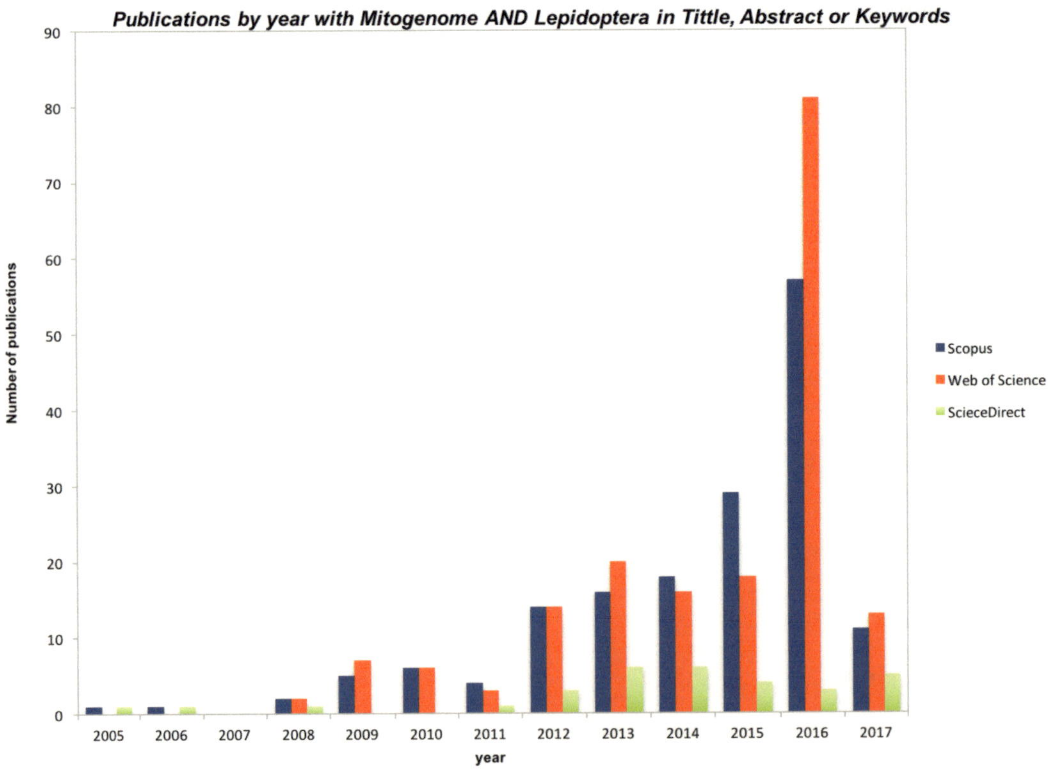

Figure 1. Growing in the number of publications with the words "Mitogenome AND Lepidoptera" at the title, abstract or keywords of scientific articles. Records were subtracted from literature databases; Scopus, Web of Science, and ScienceDirect.

approximately 500 mitogenomes of insects have been determined and subsequently deposited in GenBank [6]. Surprisingly, one of the most recent report shows that only 140 complete Lepidoptera mitogenomes (28 families from 12 superfamilies) have been sequenced and deposited in genomes databases [7], which contrasts with the number of described species in this order, as previously mentioned. In this perspective, it has been the growing research efforts of scientist around the world seeking to expand the knowledge barrier of one mitochondria of Lepidoptera. **Figure 1** shows continuous growth in the number of scientific publications in this field, showing records subtracted from literature databases: Scopus, Web of Science, and ScienceDirect.

In the present chapter, the complete mitogenome of *Tecia solanivora* is presented as a model to understand how to study a mitochondrial genome in insects. Additionally, it presented a review about mitogenomes research done mainly over lepidopteran insects considered crop pests to provide insight into aspects like genome structure and organization, nucleotide composition, codon usage, molecular functions, interactions among genes, and notable noncoding sequences included in the A + T-rich region. The utility of this study is to improve databases and support the determination of lepidopteran population genetic studies in the future.

2. Mitochondrial genome in insects

In insects, the mitochondrial genome is a circular double-stranded molecule typically between 14,000 and 20,000 bp. It contains 13 PCGs, 2 rRNAs, 22 tRNAs, and a control region (also known as the A + T-rich region), which are organized and oriented in different ways [8]. This genome has been widely used for phylogeny studies, phylogeography, population genetics, and molecular diagnostics. It has also been used to identify novel genes relevant for future studies [9], because of its small size, maternal inheritance, low recombination rate, relatively rapid evolutionary frequency, and multiple copies per cell [10]. Consequently, mitogenome sequences are rapidly evolving with about 500 insect species currently sequenced [6].

3. Characterization of insect mitogenomes

The complete mitogenome of *Tecia solanivora* is presented as a model to understand how to study and characterize a mitochondrial genome in insects. Additionally, it presented a review about mitogenomes research done mainly over lepidopteran insects considered crop pests. The argument for presenting *T. solanivora* as a model is because this lepidopteran insect represents the most damaging potato (*Solanum tuberosum*) pest in both Central and South America and Spain [11, 12]. *T. solanivora* was reported first time in Central America in 1956, affecting potato crops (*S. tuberosum*), which resulted in a direct effect on the economy. Even though this pest has a reduced mobility, it has invaded several countries in Central and South America as well as the Canary Islands in Spain where potato is grown [13]. It is important to mention that *T. solanivora* has been producing damage in both field crops and stored potato tubers, causing economic losses ranged from 50 to 100% [14]. The economic impact of the pest in countries of the Andean area is much more serious than in Central America, mainly because potato is an important family staple and its production is very intensive. Therefore, *T. solanivora* is considered the most damaging crop insect pest in such countries [15]. Nevertheless, as an insect belonging to Lepidoptera order, the study carried out on the characterization of *T. solanivora* mitogenome could be applied to other studies on insects considered pest for agriculture.

3.1. Genome sequencing and assembling

T. solanivora larvae were collected from field- or storage-infested potato tubers from Colombia. All biological samples were preserved in 70% ethanol and stored at −70°C until DNA extraction. Whole *T. solanivora* genomic DNA was extracted using the protocol described by [13]. Each sample was analyzed by electrophoresis and the DNA concentration was quantified using a NanoDrop 2000 (Thermo Scientific, Wilmington, DE, USA). The whole genome was sequenced by Illumina Hiseq 2000 system at Chapel Hill High-Throughput Sequencing Facility in the University of North Carolina. The sequencing system generated 100 bp paired-end reads with a 342 bp insert size, these reads were checked and filtered using a homemade quality criterion script and finally the paired-end reads were assembled using de novo assembler VELVET 1.2.10 [16] with an optimized k-mer parameter of 99. The mtDNA was identified through a comparison between *Tecia solanivora* scaffolds and the mtDNA sequences reported in the NCBI GenBank, resulting in the identification of a single mtDNA contig with 200× coverage. To verify the topology of the mtDNA, the reads that mapped to the borders of the contig were located at an expected distance from their respective pairs. Any discrepancies that occurred, especially in the homopolymer regions, were manually edited.

3.2. Gene annotation and compositional analysis

To predict the protein-coding genes (PCGs), rRNA genes and tRNA genes from *T. solanivora* mtDNA, their sequence was submitted to the automatic annotator of mitochondrial genes online [dual organellar genome annotation (DOGMA), http://dogma.ccbb.utexas.edu] [17]. To determinate homology between *T. solanivora* genes and other previously sequenced Lepidoptera species was used NCBI BLAST program and results manually curated. Then, the PCGs nucleotide composition, genome, and codon position were determined and the PCGs were translated into putative proteins for calculating of the relative synonymous codon usage (RSCU) using the invertebrate mitochondrial genetic code in MEGA version 5.2.2 [18]. The frequencies of A, T, G, and C were used to calculate the composition skew according to the AT- and GC- skew formulas. The intergenic and overlap sequences were pulled out manually from genome using SeqBuilder from the DNAStar package (DNAStar Inc., Madison, Wisconsin, USA).

3.3. Genome structure, organization, and base composition obtained

The *T. solanivora* mitogenome obtained was a closed circular 15,251-bp molecule (GenBank accession number KT326187) (**Figure 2**). It contains the typical set of 37 genes (13 PCGs, 22 tRNAs, and two rRNAs) and a large, 325-bp noncoding region (control region). A total of 24 genes were transcribed on the majority-coding strand (H-strand), while the rest were transcribed on the minority-coding strand (L-strand) (**Table 1**).

http://dx.doi.org/10.5772/intechopen.71158

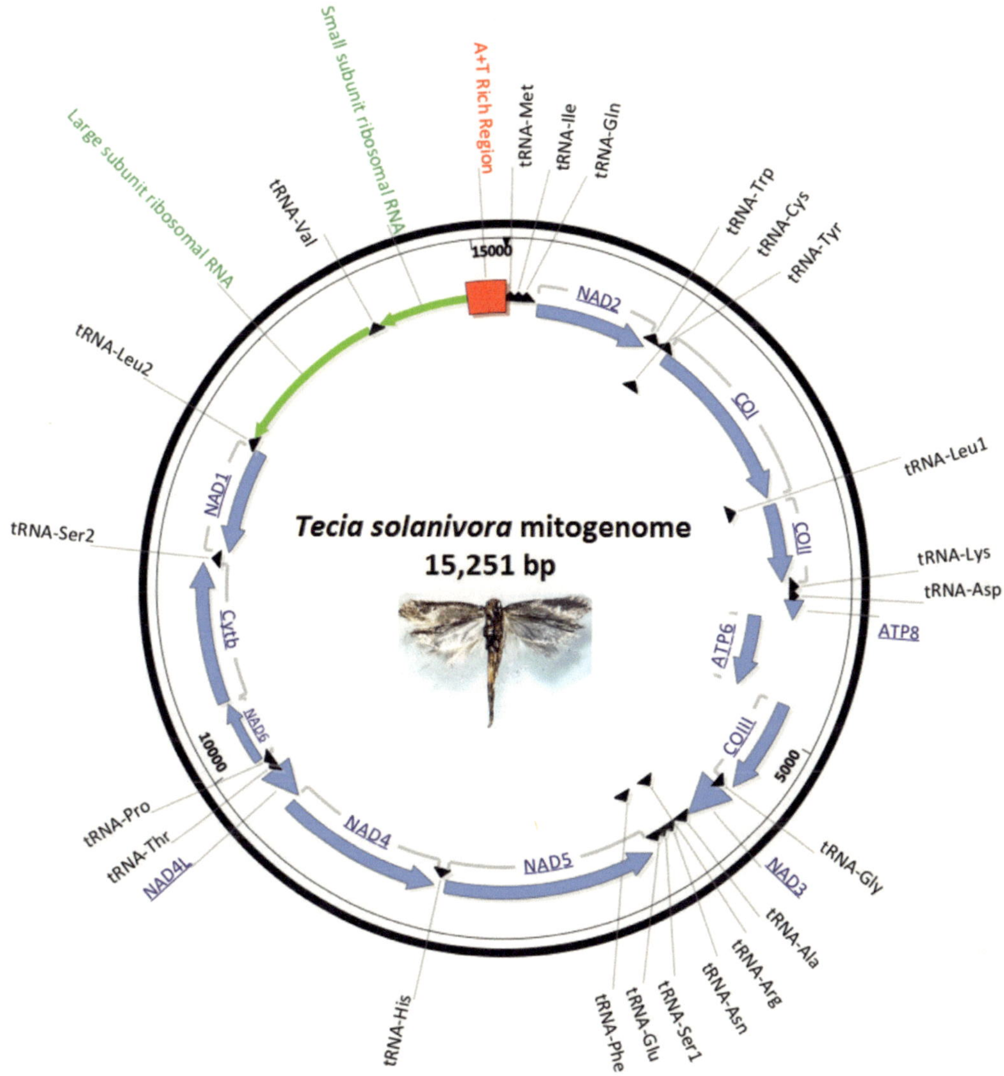

Figure 2. Map of the mitochondrial genome of T. solanivora. Protein-coding genes (names with underline) coded on the majority strand arrows going in clockwise direction, while the rest going counterclockwise. The tRNA genes are designated by tRNA-amino acid codes. The rRNAs two and they are located next to tRNA-val and the A + T-rich region (control region) is indicated by a square.

When we compared with other reported Lepidoptera family mitogenomes, it found an identical gene order and orientation of the mitochondrial genes of this species to other lepidopteran moths, including *Tryporyza incertulas* [19], *Corcyra cephalonica* [20], *Adoxophyes honmai* [21], *Apocheima cinerarius* [22], *Amata emma* [23], *Attacus atlas* [24], *Bombyx mori* [25], *Caligula boisduvalii* [26], *Chilo auricilius* [19], *Diaphania pyloalis* [27], *Manduca sexta* [9], *Ostrinia nubilalis*, *Ostrinia furnacalis* [28], *Samia Cynthia ricini* [29], and *Sasakia funebris* [30], among others.

Gene	Direction	Position (bp)	Length (bp)	Anticodon	Start codon	Stop codon
tRNA-Met	Forward	1–68	68	CAT		
tRNA-Ile	Forward	70–134	65	GAT		
tRNA-Gln	Reverse	136–204	69	TTG		
NAD2	Forward	259–1269	1011		ATT	TAA
tRNA-Trp	Forward	1268–1336	69	TCA		
tRNA-Cys	Reverse	1329–1394	66	GCA		
tRNA-Tyr	Reverse	1406–1471	66	GTA		
COI	Forward	1475–3010	1536		CGA	TAA
tRNA-Leu (UUR)	Forward	3006–3073	68	TAA		
COII	Forward	3074–3754	681		ATG	T
tRNA-Lys	Forward	3756–3826	71	CTT		
tRNA-Asp	Forward	3837–3904	68	GTC		
ATP8	Forward	3905–4072	168		ATT	TAA
ATP6	Forward	4066–4743	678		ATG	TAA
COIII	Forward	4743–5531	789		ATG	TAA
tRNA-Gly	Forward	5534–5600	67	TCC		
NAD3	Forward	5601–5954	354		ATT	TAA
tRNA-Ala	Forward	5964–6030	67	TGC		
tRNA-Arg	Forward	6030–6095	66	TCG		
tRNA-Asn	Forward	6101–6166	66	GTT		
tRNA-Ser (AGN)	Forward	6181–6246	66	GCT		
tRNA-Glu	Forward	6247–6315	69	TTC		
tRNA-Phe	Reverse	6314–6380	67	GAA		
NAD5	Reverse	6364–8097	1734		ATT	TAA
tRNA-His	Reverse	8113–8178	66	GTG		
NAD4	Reverse	8183–9523	1341		ATG	TAA
NAD4L	Reverse	9523–9816	294		ATG	TAA
tRNA-Thr	Forward	9819–9883	65	TGT		
tRNA-Pro	Reverse	9884–9949	66	TGG		
NAD6	Forward	9952–10,479	525		ATA	TAA
Cytb	Forward	10,497–11,642	1146		ATA	TAA
tRNA-Ser	Forward	11,646–11,712	67	TGA		
NAD1	Reverse	11,730–12,665	936		ATA	TAG
tRNA-Leu	Reverse	12,669–12,736	68	TAG		

Gene	Direction	Position (bp)	Length (bp)	Anticodon	Start codon	Stop codon
rRNA-Large	Reverse	12,737–14,065	1329			
tRNA-Val	Reverse	14,089–14,155	67	TAC		
rRNA-Small	Reverse	14,157–14,926	770			
A + T region		14,927–15,251	325			

Table 1. Summary of *T. solanivora* mitogenome.

The typical lepidopteran arrangement of the tRNAs (tRNA-Met, tRNA-Ile, tRNA-Gln) was observed in the *T. solanivora* mitogenome but differs from the order found in ancient insects (**Figure 3**). In that sense, it was determined that *T. solanivora* presents several differences from the ancestral organization of the tRNA-Met region (A + T-rich region, tRNA-Ile, tRNA-Gln, tRNA-Met) [20], which is also found in the mitogenomes of *Aedes aegypti* (Diptera) [31] and *Acrida cinerea* (Orthoptera) [32]. In the case of *T. solanivora*, the order is: A + T region, tRNA-Met, tRNA-Ile, tRNA-Gln. Additionally, in the *T. solanivora* mitogenome, the tRNA-Lys gene is found after the COII gene, contrary to *A. cinerea*, where they are found in the reverse order. In addition, the NAD3 gene was located before the tRNA-Ala gene in *T. solanivora*, whereas in *A. aegypti*, the gene located in this region is tRNA-Arg.

The nucleotide composition determined in the entire *T. solanivora* mitogenome was A: 38.6, T: 39.6, C: 13.3, and G: 8.4% (**Table 2**). This nucleotide composition shows that highly A + T-biased (78.2%) with a similar proportion of adenine (A) and thymine (T) compared with the reported ranges found in other Lepidoptera mitogenomes. In the same way, *T. solanivora* mitogenome exhibits negative AT-skew (−0.013) and GC-skew (−0.226) values (**Table 3**). However, the most of Lepidoptera family members have shown higher percentages of A than T, such as: *O. nubilalis* (A: 41.3, T: 38.8%), *O. furnacalis* (A: 41.46, T: 38.92%), *B. mori* (A: 43.06, T: 38.30%) [33], *Phthonandria atrilineata* (A: 40.78, T: 40.24%) [34], *Ochrogaster lunifer* (A: 40.09, T: 37.75%) [35], *Chinese Bombyx mandarina* (A: 43.11, T: 38.48%) [36], and *A. atlas* (A: 39.8, T: 39.5%), among others. Likewise, the cytosine (C) content in the *T. solanivora* mitogenome was greater than guanine (G), which is similar to the percentages identified in other recently discovered Lepidoptera mitogenomes, except for *Antheraea yamamai* (G: 10.71, C: 10.35%) [37], *Eriogyna pyretorum* (G: 10.61, C: 7.45), and *Artogeia melete* (G: 11.33, C: 8.65%) [38].

3.4. Protein-coding genes (PCGs)

The protein-coding genes (PCGs) encompassed 11,191 bp of the entire assembled sequence (73.38%) and exhibited an A + T content of 76.4%. Nine of the 13 PCGs are coded on the majority strand (ATP6, ATP8, COI, COII, COIII, Cytb, NAD2, NAD3, and NAD6), while the rest (NAD1, NAD4, NAD4L, and NAD5) are coded on the minority strand. For the protein-coding genes, the A + T content was calculated for the three-codon positions, and they showed few differences from other Lepidoptera mitogenomes. *T. solanivora* showed negative AT- and GC-skew values in the second and third positions of each codon, indicating a greater inclination for the nitrogen bases, T and C, while the first position showed a slightly positive value

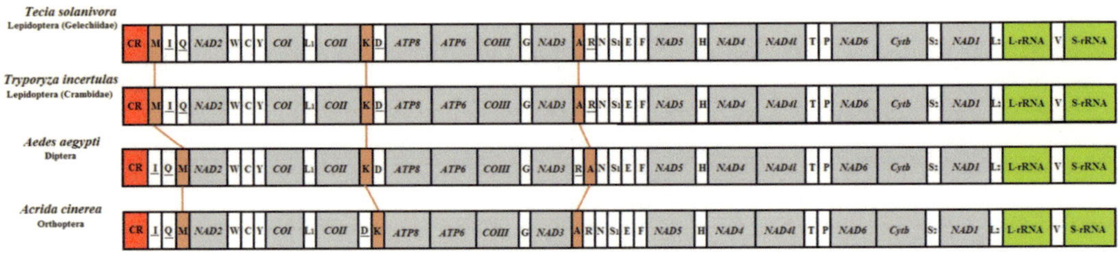

Figure 3. Gene arrangement of the *T. solanivora* mitogenome. Protein-coding genes are marked by light gray, ribosomal RNA genes by light green, control region by red and tRNA genes are designated by the single letter amino acid code (white). Brown box and horizontal line represent gene clusters that changed positions.

nt %	Whole mtDNA	Protein-coding sequence			Concatenated PCGs	rRNAs	tRNAs	IGs	A + T-rich region
		1st#	2nd#	3rd#					
A%	38.6	35.3	21.3	38.3	31.7	43.8	40.5	44.4	42.8
T%	39.6	36.6	48.2	49.2	44.7	39.9	40.2	44.4	48.3
C%	13.3	11.0	17.0	7.3	11.7	5.2	8.1	7.6	6.2
G%	8.4	17.7	13.5	5.2	11.9	11.1	11.2	3.5	2.8
A + T%	78.2	71.9	69.5	87.5	76.4	83.7	80.7	88.8	91.1
C + G%	21.7	28.7	30.5	12.5	23.7	16.3	19.3	11.1	9.0
AT-Skew%	-0.013	-0.018	-0.387	-0.125	-0.170	0.047	0.004	0	-0.060
GC-Skew%	-0.226	0.233	-0.115	-0.168	0.008	0.362	0.161	-0.36	-0.378

Table 2. Nucleotide composition of *T. solanivora* mitogenome.

Species	Length (bp)	A%	G%	T%	C%	A + T%	G + C%	AT-skew	GC-skew
T. solanivora	15,251	38.6	8.4	39.6	13.3	78.2	21.7	-0.013	-0.226
A. selene	15,236	38.54	8.05	40.37	13.03	78.91	21.08	-0.023	-0.236
C. raphaelis	15,314	39.37	7.30	43.29	10.04	82.66	17.34	-0.047	-0.158
E. pyretorum	15,327	39.17	7.63	41.65	11.55	80.82	19.18	-0.031	-0.204
A. yamamai	15,338	39.26	7.69	41.04	12.02	80.30	19.71	-0.022	-0.220
C. boisduvalii	15,360	39.34	7.58	41.28	11.79	80.62	19.37	-0.024	-0.217
S. cynthia ricini	15,384	39.65	7.81	40.13	12.41	79.78	20.22	-0.006	-0.227
M. sexta	15,516	40.67	7.46	41.11	10.76	81.78	18.22	-0.005	-0.181
A. pernyi	15,566	39.22	7.77	40.94	12.07	80.16	19.84	-0.021	-0.217
A. honmai	15,680	40.15	7.88	40.24	11.73	80.39	19.61	-0.001	-0.196

Table 3. Comparison of nucleotide composition and skewness between *T. solanivora* and other lepidopteran mitogenomes.

for the GC-skew indicating a greater bias for G than C (**Table 3**). In general, the codons of *T. solanivora* mitogenome present high A + T content for the first position (71.9%). Similar values were observed in *A. emma* (73.1%), *Antheraea pernyi* (72.9%) [39], *C. boisduvalii* (73.8%), *B. mandarina* (75.0%) [33], and *M. sexta* (74.8%), but in the second position was found a lower A + T percentage (69.5%).

Twelve PCGs were identified in the *T. solanivora* mitogenome with the typical ATN initiation codons (isoleucine and methionine), except for the COI gene that is initiated by CGA initiation codon (arginine). The typical ATN codon represents a putative codon commonly observed in the order Lepidoptera [9, 40], and is thus considered a synapomorphy of this group of insects [41]. The methionine start codon, ATG, was used by five of the 13 PCGs, and ATA was used to initiate protein synthesis in the *NAD1, NAD6,* and *cytb* genes. In contrast, an atypical isoleucine codon (ATT) was used to initiate protein synthesis in the *ATP8, NAD2, NAD3,* and *NAD5* genes. Arginine (CGA) was used for the COI gene for which a nucleotide sequence of four or six base pairs has been proposed, much like TTAG in *Maruca vitrata* [42], to serve in a nonstandard initiation process located immediately upstream from the putative arginine CGA start codon of COI. In *T. solanivora,* this tetranucleotide sequence consisted of TTGG. The high A + T percentages in insect mitogenomes result in high probabilities of finding a noncoding triplet or a coding triplet within the tRNA-Tyr gene, a result that could potentially produce generalized annotation errors for the gene COI [43]. Previous studies have discussed the possibility that translation initiation of this gene involves an unusual sequence of four to six nucleotides (ATAA, TTAA, GTAA, ATTA, or ATTTAA) located immediately before the coding primer of the COI gene. This sequence apparently functions as the translation initiator in the majority of insects from the family Diptera, including *Drosophila yakuba* [44]. However, in *T. solanivora,* the sequence TTGG was found immediately before putative initiation codon CGA.

For the stop codon genes, we found the TAA codon in 11 of the PCGs, coinciding with the mitogenomes of other Lepidoptera, including *T. incertulas, S. funebris, S. cynthia, and A. emma* of the family Hesperiidae. In the *NAD1* gene, the TAG stop codon was found and similar results were obtained in five species of the family Hesperiidae [41], while the *COII* gene used a single T as an incomplete stop codon, which is commonly found in the majority of Lepidoptera species to date [9, 41]. This truncated codon could be a representative of a recognition site for an endonuclease that splits the polycistronic pre-mRNA, where a post-transcriptional polyadenylation then occurs, resulting in a functional stop codon (TAA) [9, 19, 45].

The CDpT or Codons Per Thousand Codons of the *T. solanivora* mitogenome was calculated, and five amino acid families were identified. The most common families were: phenylalanine (Phe), asparagine (Asn), isoleucine (Ille), lysine (Lys), and leucine 2 (Leu 2), as shown in (**Figure 4a**), being in *T. solanivora,* phenylalanine (Phe) the most abundant, instead of Leucine 2 (Leu2), which dominates in other Lepidoptera mitogenomes [35]. Additionally, when the relative synonymous codon use (RSCU) was determined, we identified that *T. solanivora* presents all typical codons found in other invertebrates (**Figure 4b**). The codons were richer in A or T at the third position and consequently have less G or C.

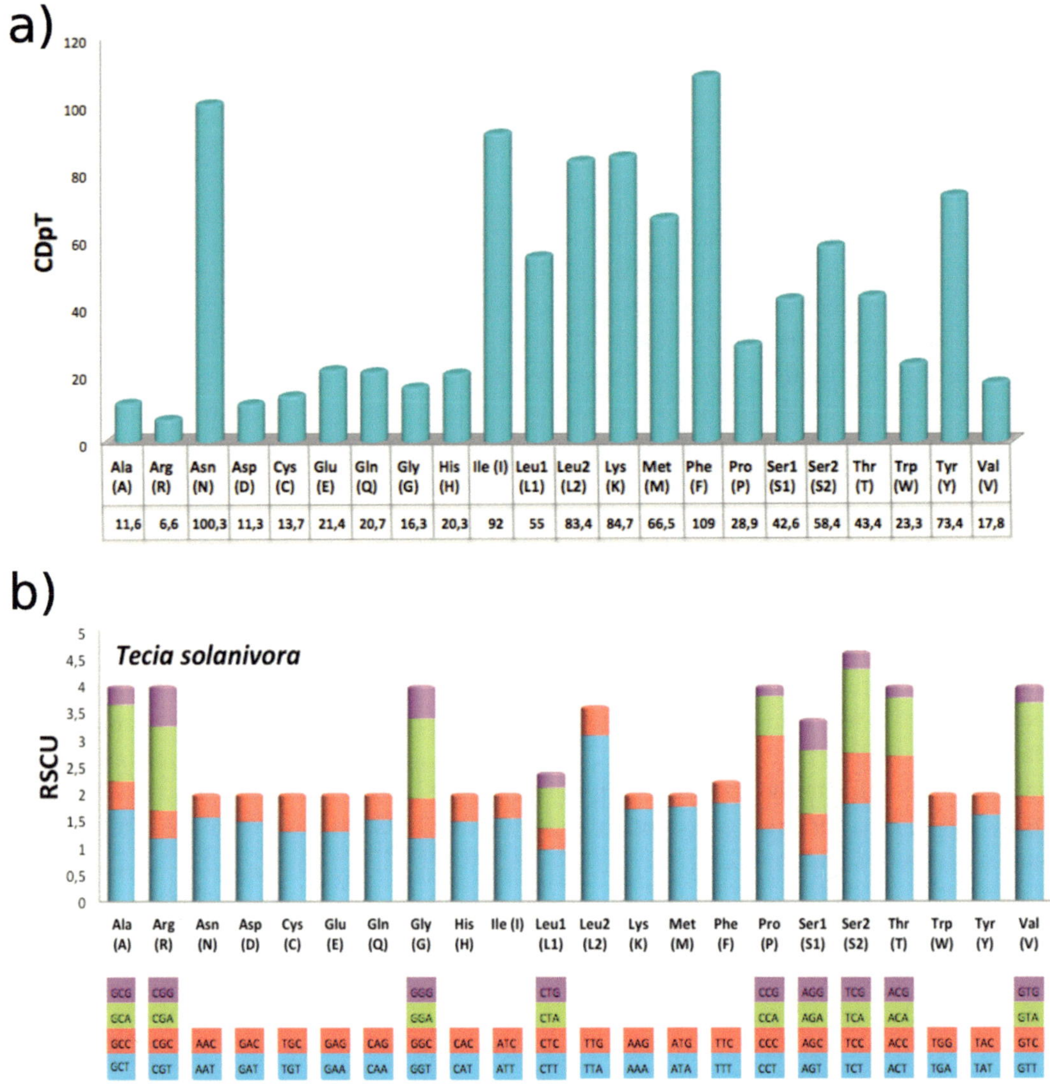

Figure 4. Codon distribution and relative synonymous codon usage (RSCU) in *T. solanivora* mitogenome. (a) Codon distribution and (b) RSCU. Codon families are provided on the X axis and the RSCU on the Y axis. This mitogenome presents all possible codon families existing in Lepidoptera.

3.5. Transfer RNA and ribosomal RNA genes

It was found that *T. solanivora* contains a typical set of 22 tRNAs with a high A + T bias, accounting for 80.7% of the tRNAs, slightly positive AT-skew (0.004) and a clearly positive GC-skew (0.161). These results suggest that tRNAs exhibit a higher inclination for nitrogen bases A and G than for T and C. Similar results were reported by [46] in *E. pyretorum* (AT-skew = 0.039 and GC-skew = 0.174) and [38] in *A. melete* (AT-skew = 0.034 and GC-skew = 0.142). Among the tRNA genes, 14 are coded on the H-strand and eight on the L-strand with lengths ranged between 65 and 71 bp, and these genes exhibited positive AT-skew (0.047) as well as in almost all lepidopteran mitogenomes. However, the rearrangement of certain tRNAs were found translocated in the *T. solanivora* mitogenome compared with out-groups, which are described in (**Figure 3**).

Similar to other mitochondrial sequences from insect species, there were two rRNAs in *T. solanivora* with a total length of 2099 bp and an AT content of 83.7% (**Table 2**). The large ribosomal gene (rRNA-Large), located between tRNA-Leu1 and tRNA-Val, has a length of 1329 bp, whereas the small gene (rRNA-Small), located between tRNA-Val and the A + T-rich region, has a length of 770 bp (**Table 1**). These rRNA lengths were within the range of values reported for other Lepidoptera, as their values range between 1314 bp in *Euploea mulciber* (Nymphalidae) [47] and 1330 bp in *Coreana raphaelis* (Lycaenidae) [48]. For the rRNA-Large and rRNA-Small, the length ranges from 739 bp for *Protantigius superans* (Lycaenidae) to 788 bp in (Nymphalidae) [49]. The rRNAs in *T. solanivora* have an A + T content of 83.7%, and similar values were reported for other Lepidoptera, including *C. cephalonica* (80.43%), *T. incertulas* (82.8%), and *Dichocrocis punctiferalis* (85.1%) [50].

3.6. Noncoding and overlapping regions

Most of the intergenic regions in this mitogenome were short (≤15 bp) and the total length of the noncoding regions in the mtDNA of *T. solanivora* was 199 bp. This region is composed by 21 intergenic spacer sequences, ranging from 1 to 54 bp and showed highly A + T-biased (88.8%) (**Table 2**). The intergenic spacers longer were denominated S1, S2, S3, and S4. Intergenic sequence S1 is commonly found in Lepidoptera mitogenome order between the tRNA-Gln and NAD2 genes and length ranges between 38 pb in *T. incertulas* and 88 bp in *Sasakia charonda* [50]. However, this region has not been identified in insects that belong to other orders [9].

This sequence (S1) could be considered as a mitogenome marker for Lepidoptera order, and it most likely originated from a partial NAD2 gene duplication [19]. Intergenic sequence S2 (23 bp) was found between rRNA-Large and tRNA-Val. Intergenic sequences S3 and S4 (17 bp) separate genes NAD6 and Cytb, and the tRNA-Ser2 and NAD1 genes, respectively. The latter sequence contains the "ATACTAA" motif, typically found in other lepidopterans [9, 23, 51]. This motif plays an apparent role as a recognition site for the protein implicated in mitochondrial transcription termination (mtTERM) [52]. Furthermore, this sequence has been recognized for being highly conserved, with a length ranging between 17 and 20 bp [23].

Furthermore, in the *T. solanivora* mitogenome, three principal overlap sequences were identified and were designated as OLS1, OLS2, and OLS3. OLS1 was found overlapping the tRNA-Phe and NAD5 genes. This sequence presents the greatest length, with a total of 17 bp. OLS2 was found between the tRNA-Trp and tRNA-Cys genes, with a total length of 8 bp and the 7-bp OLS3-overlapped genes ATP8 and ATP6, which are consistent with the same genes found in other lepidopterans, although they differ in length [37, 52]. In addition, an unusual overlap region (OLS1) was found between the tRNA-Phe and NAD5 genes; it is important to mention that this 17-bp region has not been reported before.

3.7. The A + T-rich region

The A + T-rich region is a noncoding region with 325 bp length located between rRNA-Small and tRNA-Met. The region contains 91.1% AT nucleotides, with negative AT- and GC-skew values (**Table 2**), meaning that it is biased for the nitrogen base thymine, as reported for the mitogenomes of other lepidopterans. One exception to this trend is *A. honmai*, which has a positive AT skew (0.028), indicating a bias for adenines [21]. The length of this region is variable in the other Lepidoptera, and it can be as long as 1270 bp, as reported in *Papilio bianor* (Papilionidae) [53].

This A + T region is a conserved structure commonly found in other Lepidoptera, which includes the "ATAGA" motif followed by a 17-bp poly-T stretch, just like in *T. solanivora* mitogenome. This motif is immediately followed by the tRNA-Met gene [35, 41, 54] and it seems it has an important role in the replication initiation in minor strand of mtDNA in addition to gene regulation [19, 30, 41]. Furthermore, eight microsatellite regions were identified within the mitogenome of *T. solanivora*, referred to as $(TAA)_4$, $(AT)_8$, and $(TAT)_7$. These were the most representative microsatellites found in the species, although the mononucleotide sequences, $(T)_6$ and $(A)_{10}$, were also identified [55]. These represent the relevant regions of this genome for future studies. Also, in most lepidopteran mitogenomes, the $(AT)_8$ microsatellite has been previously reported. This microsatellite is preceded by the "ATTTA" motif that is commonly found in other mitogenomes [9, 41]. In *S. funebris*, the same $(AT)_8$ microsatellite was identified as that found in *T. solanivora*. Nevertheless, most lepidopteran mitogenomes report $(AT)_n$, where n ranges from 7 to 12 [19, 23].

3.8. Phylogenetic relationships

To illustrate the phylogenetic relationship of *T. solanivora* (Lepidoptera: Gelechiidae) with other 16 Lepidoptera families, we used a concatenated set of PCGs of 72 other complete Lepidoptera mitogenomes obtained from GenBank with previous elimination of start and stop codons. The phylogenetic relationship among the eight Lepidoptera superfamilies was inferred using both Bayesian Inference and maximum likelihood methods, which produced similar topologies to previously analyzed phylogenies obtained for other lepidopterans. The results obtained with both methods produced similar and consistent topologies. Our results showed high support values for the majority of the nodes and thus the interrelationships are well-resolved within order Lepidoptera. We used *Aedes aegypti* (Diptera) and *Acrida cinerea* (Orthoptera) as out-groups, the phylogenetic trees revealed nine Lepidoptera clades. Species of the Papilionoidea, Noctuidae, Bombycidae, Geometridae, Pyralidae, Gelechiidae, Tortricidae, Yponomeutidae, and Hepialidae superfamilies cluster monophyletic groups, with strongly supported bootstrapped and posterior probabilities (100). All those results and analysis were published by authors of the present chapter in 2016 [7].

4. Other recent studies with mitogenomes of Lepidopteran considered crop pests

A search carried out on August 04, 2017 in Scopus database showed that after publishing the scientific paper made by authors from this chapter [7], scientists have published 62 other studies on mitochondrial genomes of lepidopteran insects. Most of them were focused on understanding their composition, organization, motifs, and the inference of phylogenetic relationships between these organisms [41, 56–58]. However, recently, [59] reported besides of typifying the mitogenomes of *Mesophleps albilinella* and *Dichomeris ustalella* (Lepidoptera: Gelechiidae), the prediction of the secondary structures of the tRNAs [44, 60]. In this model, single polynucleotide chain form four or five arms to fold itself with each other, like a clover leaf, call them an acceptor arm, DHU or D arm, anti-codon arm, TψC arm, and variable arm [61], which play a role in proper folding of the tRNA into the L-shaped tertiary structure while modifications in or around the anti-codon loop contribute to the function of tRNAs in decoding [62]. Prediction of secondary structures of tRNAs has been using tRNAscan-SE 1.21, Mito/Chloroplast, invertebrate genetic code for the prediction of tRNA isotypes, and a cutoff of 1 score [63]. This method allowed to find and predict the secondary structure of 21 tRNAs in both species, except for the tRNASer (AGN), which has a truncated DHU arm, with consideration given to the anti-codons [59]. Frequently, have been reported tRNA-like structures into the A + T-rich region in Lepidoptera [37, 49, 64], but it seems to be fake tRNAs of random secondary structures, owing to the reduced sequence complexity (>90% A + T) in this noncoding region [59], suggesting that are fake tRNAs of random secondary structures, owing to the reduced sequence complexity (>90% A + T) in this noncoding region [65].

On the other hand, we must highlight the importance of studying of insect pests mitogenomes, this allows to propose hypotheses related with the evolutionary origin of the different larval stages, which causes significant damage to crops during this state, and could predict which is the most adaptable state to each type of environment as for example in Parapoynx *crisonalis* moth (Lepidoptera: Crambidae) [66] in which its larvae are lacking tracheal gills because they are pest in aquatic crops [67]. This analysis also can be extrapolated to the study of moths that are also plagues only in larval state but in terrestrial crops such as *T. solanivora*, and in this way to be able to find the most viable way to control this type of pests.

5. Novel techniques for pest control using mtDNA

Pest species represent a major ongoing threat to global biodiversity, demanding effective management approaches are required that regulate pest numbers, while minimizing collateral damage to nontarget species. Species-specific pest controls have been developed in order to be long-lasting measures and effectives [68]. One of these methods is called the sterile insect technique (SIT), whereby sterile males are introduced into target populations, so that they could be produced continuously within the targeted populations for control, and thus reducing production of females when mating with them. However, the SIT generally requires continuous large-scale production and introduction of sterile evils to sustain population suppression [69].

At the level of maternally inherited mitochondrial DNA (mtDNA) has been identified naturally occurring mutations that cause male infertility. These mutations have little or no impact on females, and hence are minimally or not selected against (i.e. are self-perpetuating in nature). Due to those kinds of mutations, have only been identified in some model systems such as mice and fruit flies, they are likely to be widespread in nature threatening small populations viability of endangered species. Currently, a novel variant of the SIT, is the recently proposed Trojan female technique (TFT), based on the use of naturally occurring mutations or induced by CRISPR-Cas9 (clustered, regularly interspaced, short palindromic repeats system) in the mtDNA [69]. The consortium aims to harness these mutations to develop a widely applicable capability for pest control, through the release of Trojan females carrying the mutations [68, 69].

With this technique, males that inherit these mutations will have fewer offspring than wild-type males, while females will remain normal (fertile). It is well known that mtDNA is generally maternally inherited, so this sex-bias in effects will reduce selection pressure against the TFT mutation. When females carrying the TFT mutation are released into a pest population, they could cause multi-generational population suppression. However, while promising well and scientific means to control pest populations or disease vectors, the release of genetically engineered animals raises into ethical issues and a debate is currently underway discussing safety and regulatory concerns [68, 69].

6. Conclusion

In this chapter, the complete mitochondrial genome of *T. solanivora* was presented as a model to understand how to characterize and study a mitogenome in insects. It was sequenced, analyzed, and compared with other lepidopteran insects. This mitogenome shares many features with those reported previously in Lepidoptera but exhibited several subtle differences in the codon distribution within the A + T region. The phylogenetic relationships of nine clades of the order Lepidoptera were developed using Bayesian and maximum likelihood inference, which provided well-supported results compared with other phylogenies based on both molecular and morphological traits. In addition, an update about other recent mitogenomes research done mainly over lepidopteran insects considered crop pests was made. On the other hand, it was shown a novel development based on induced mutations by CRISPR-Cas9 in the mitogenomes seeking applicable capability for pest control. The utility of all information presented in this chapter is to improve scientific databases and support the determination of lepidopteran population genetic studies in the future.

Author details

Viviana Ramírez-Ríos, Javier Correa Alvarez and Diego Villanueva-Mejia*

*Address all correspondence to: dvillanu@eafit.edu.co

Department of Biological Sciences, School of Sciences, Universidad EAFIT,
Medellin, Antioquia, Colombia

References

[1] Nutter FW, Teng PS, Royer MH. Terms and concepts for yield, crop loss, and disease thresholds. Plant Disease. 1993;**77**:211-215

[2] Cerda R, Avelino J, Gary C, et al. Primary and secondary yield losses caused by pests and diseases: Assessment and modeling in coffee. PLoS One. 2017;**12**:e0169133

[3] EC O. Crop losses to pests. The Journal of Agricultural Science. 2006;**144**:31

[4] Kristensen NP, Scoble MJ, Karsholt O. Lepidoptera phylogeny and systematics: The state of inventorying moth and butterfly diversity. Zootaxa. 2007;**1668**:699-747

[5] Ma H-F, Zheng X-X, Peng M-H, et al. Complete mitochondrial genome of the meadow moth, *Loxostege sticticalis* (Lepidoptera: Pyraloidea: Crambidae), compared to other *Pyraloidea moths*. Journal of Asia-Pacific Entomology. 2016;**19**:697-706

[6] Cameron SL. How to sequence and annotate insect mitochondrial genomes for systematic and comparative genomics research. Systematic Entomology. 2014;**39**:400-411

[7] Ramírez-Ríos V, Franco-Sierra ND, Alvarez JC, et al. Mitochondrial genome characterization of *Tecia solanivora* (Lepidoptera: Gelechiidae) and its phylogenetic relationship with other lepidopteran insects. Gene. 2016;**581**:107-116

[8] Boore JL. Animal mitochondrial genomes. Nucleic Acids Research. 1999;**27**:1767-1780

[9] Cameron SL, Whiting MF. The complete mitochondrial genome of the tobacco hornworm, *Manduca sexta,* (Insecta: Lepidoptera: Sphingidae), and an examination of mitochondrial gene variability within butterflies and moths. Gene. 2008;**408**:112-123

[10] Avise JC. Molecular Markers, Natural History and Evolution. New York: Chapman & Hall; 1994

[11] Pulliandre N, Dupas S, Dangles O, Zeddam J, Capdevielle-Dulac C, Barbin K, Torres-Leguizamon M, Silvain J. Genetic bottleneck in invasive species: The potato tuber moth adds to the list. Biological Invasions. 2008;**10**:319-333

[12] Torres-Leguizamón M, Dupas S, Dardon D, Gómez Y, Niño L, Carnero A, Padilla A, Merlin I, Fossoud A, Zeddam J, Lery X, Capdevielle-Dulac C, Dangles O, Silvain J. Inferring native range and invasion scenarios with mitochondrial DNA: The case of *T. solanivora* successive north-south step-wise introductions across Central and South America. Biological Invations. 2011;**13**:1505-1519

[13] Villanueva-Mejía DF, Ramírez-Ríos V, Arango-Lsaza RE, et al. Phylogeography of *Tecia solanivora* from Colombia based on cytochrome oxydase i and cytochrome b mitochondrial genes. South West Entomology. 2015;**40**:775-788

[14] Zeddam JL, Orbe K, Léry X, Dangles O, Dupas S, Silvain J. An isometric virus of the potato tuber moth *Tecia solanivora* (Povolny) (Lepidoptera: Gelechiidae) has a tri-segmented RNA genome. Journal of Invertebrate Pathology. 2008;**99**:204-211

[15] Villanueva D, Saldamando CI. *Tecia solanivora*, Povolny (Lepidoptera: Gelechiidae): una revisión sobre su origen, dispersión y estrategias de control biológic. Ing y Cienc. 2013;**9**:197-214

[16] Daniel ZR, Ewan B. Velvet: Algorithms for de novo short read assembly using de Bruijn graphs. Genome Research. 2008;**18**:821-829

[17] Wyman Stacia K, Jansen RK, Bore JL. Automatic annotation of organellar genomes with DOGMA. Bioinformatics. 2004;**20**:3252-3255

[18] Koichiro T, Daniel P, Nicholas P, Glen S, Nei M, Kumar S. MEGA5: Molecular evolutionary genetics analysis using maximum likelihood, evolutionary distance, and maximum parsimony methods. Molecular Biology and Evolution. 2011:1-9

[19] Cao S-S, Du Y-Z. Characterization of the complete mitochondrial genome of *Chilo auricilius* and comparison with three other rice stem borers. Gene. 2014;**548**:270-276

[20] YP W, Li J, Zhao JL, TJ S, Luo AR, Fan RJ, Chen MC, Wu CS, Zhu C. The complete mitochondrial genome of the rice moth, *Corcyra cephalonica*. Journal of Insect Science. 2012;**12**:1536-2442

[21] Eun-Seung L, Sik SK, Moo-Sang K, Hyungjin P, Cho Soowon KC-B. The mitochondrial genome of the smaller tea tortrix *Adoxophyes honmai* (Lepidoptera: Tortricidae). Gene. 2006;**373**:52-57

[22] Liu S, Xue D, Cheng R, Han H. The complete mitogenome of *Apocheima cinerarius* (Lepidoptera: Geometridae: Ennominae) and comparison with that of other lepidopteran insects. Gene. 2014;**547**:136-144

[23] Lu H-F, Su T-J, Luo A-R, Zhu C-D, Wu C-S. Characterization of the complete mitochondrion genome of diurnal moth *Amata emma* (Butler) (Lepidoptera: Erebidae) and its phylogenetic implications. PLoS One. 2013;**8**:1-14

[24] Chen MM, Li Y, Chen M, Wang H, Li Q, Xia RX, Zeng CY, Li Y. Complete mitochondrial genome of the atlas moth, *Attacus atlas* (Lepidoptera: Saturniidae) and the phylogenetic relationship of Saturniidae species. Gene. 2014;**545**:95-101

[25] Dai LS, Zhu BJ, Liu QN, Wei GQ, Liu CL. Characterization of the complete mitochondrial genome of *Bombyx mori* strain H9 (Lepidoptera: Bombycidae). Gene. 2013;**519**:326-334

[26] Hong MY, Lee EM, Jo YH, et al. Complete nucleotide sequence and organization of the mitogenome of the silk moth *Caligula boisduvalii* (Lepidoptera: Saturniidae) and comparison with other lepidopteran insects. Gene. 2008;**413**:49-57

[27] Zhu BJ, Liu QN, Dai LS, Wang L, Sun Y, Lin KZ, Wei GQ, Liu CL. Characterization of the complete mitochondrial genome of *Diaphania pyloalis* (Lepidoptera: Pyralididae). Gene. 2013;**527**:283-291

[28] Coates BS, Sumerford DV, Hellmich RL, Lewis LC. Partial mitochondrial genome sequences of *Ostrinia nubilalis* and *Ostrinia furnicalis*. International Journal of Biological Sciences. 2005;**1**:13-18

[29] Kim JS, Park JS, Kim MJ, Kang PD, Kim SG, Jin BR, Han YS, Kim I. Complete nucleotide sequence and organization of the mitochondrial genome of eri-silkworm, *Samia cynthia ricini* (Lepidoptera: Saturniidae). Journal of Asia-Pacific Entomology. 2012;**15**:162-173

[30] Wang JP, Cao TW, Xuan SB, Wang H, Zhang M, bo Ma E. The complete mitochondrial genome of *Sasakia funebris* (Leech) (Lepidoptera: Nymphalidae) and comparison with other Apaturinae insects. Gene. 2013;**526**:336-343

[31] Behura SK, Lobo NF, Haas B, deBruyn B, Lovin DD, Shumway MF, Puiu D, Romero-Severson J, Nene V, Severson D. Complete sequences of mitochondria genomes of *Aedes aegypti* and *Culex quinquefasciatus* and comparative analysis of mitochondrial DNA fragments inserted in the nuclear genomes. Insect Biochemistry and Molecular Biology. 2011;**41**:770-777

[32] Nian L, Yuan H. Complete mitochondrial genome sequence of *Acrida cinerea* (Acrididae: Orthoptera) and comparative analysis of mitochondrial genomes in orthoptera. Comparative and Functional Genomics. 2010:1-16

[33] Yukuhiro K, Sezutsu H, Itoh M, Shimizu K, Banno Y. Significant levels of sequence divergence and gene rearrangements have occurred between the mitochondrial genomes of the wild mulberry silkmoth, *Bombyx mandarina*, and its close relative, the domesticated silkmoth, *Bombyx mori*. Molecular Biology and Evolution. 2002;**19**:385-1389

[34] Yang L, Wei ZJ, Hong GY, Jiang ST, Wen L. The complete nucleotide sequence of the mitochondrial genome of *Phthonandria atrilineata* (Lepidoptera: Geometridae). Molecular Biology Reports. 2009;**36**:1441-1449

[35] Paola S, Simonato Mauro BA, Negrisolo E. The complete mitochondrial genome of the bag-shelter moth *Ochrogaster lunifer* (Lepidoptera, Notodontidae). BMC Genomics. 2008;**9**:331

[36] Pan M, Yu Q, Xia Y, et al. Characterization of mitochondrial genome of Chinese wild mulberry silkworm, *Bomyx mandarina* (Lepidoptera: Bombycidae). Science in China. Series C, Life Sciences. 2008;**51**:693-701

[37] Kim MI, Baek JY, Kim MJ, Jeong HC, Kim KG, Bae CH, Han YS, Jin BR, Kim I. Complete nucleotide sequence and organization of the mitogenome of the red- spotted apollo butterfly, *Parnassius bremeri* (Lepidoptera: Papilionidae) and comparison with other lepidopteran insects. Molecules and Cells. 2009;**28**:347-363

[38] Hong G, Jiang S, Yu M, Yang Y, Li F, Xue F, Wei Z. The complete nucleotide sequence of the mitochondrial genome of the cabbage butterfly, *Artogeia melete* (Lepidoptera: Pieridae). Acta Biochimica et Biophysica Sinica Shanghai. 2009;**41**:446-455

[39] Liu Y, Li Y, Pan M, Dai F, Zhu X, Lu C, Xiang Z. The complete mitochondrial genome of the Chinese oak silkmoth, *Antheraea pernyi* (Lepidoptera: Saturniidae). Acta Biochimica et Biophysica Sinica Shanghai. 2008;**40**:693-703

[40] Fenn JD, Cameron SL, Whiting M. The complete mitochondrial genome sequence of the *Mormon cricket* (Anabrus simplex: Tettigoniidae: Orthoptera) and an analysis of control region variability. Insect Molecular Biology. 2007;**16**:239-252

[41] Kim MJ, Wang AR, Park JS, et al. Complete mitochondrial genomes of five skippers (Lepidoptera: Hesperiidae) and phylogenetic reconstruction of Lepidoptera. 2014;**549**: 97-112.

[42] Margam VM, Coates BS, Hellmich RL, Agunbiade T, Seufferheld MJ, et al. Mitochondrial genome sequence and expression profiling for the legume pod borer *Maruca vitrata* (Lepidoptera: Crambidae). PLoS One;**6**. Epub ahead of print 2011. DOI: 10.1371/journal.pone.0016444

[43] Stewart JB, Beckenbach A. Characterization of mature mitochondrial transcripts in Drosophila, and the implications for the tRNA punctuation model in arthropods. Gene. 2009;**445**:49-57

[44] Clary DO, Wolstenholme DR. Genes for cytochrome c oxidase subunit I, URF2, and three tRNAs in Drosophila mitochondrial DNA. Nucleic Acids Research. 1983;**11**:6859-6872

[45] Ojala D, Montoya J, Attardi G. tRNA punctuation model of RNA processing in human mitochondria. Nature. 1981;**290**:470-474

[46] Jiang ST, Hong GY, Yu M, Li N, Yang Y, Liu YQ, Wei ZJ. Characterization of the complete mitochondrial genome of the giant silkworm moth, *Eriogyna pyretorum* (Lepidoptera: Saturniidae). International Journal of Biological Sciences. 2009;**5**:351-365

[47] Hao J, Sun M, Shi Q, Sun X, Shao L, Yang Q. Complete mitogenomes of *Euploea mulciber* (Nymphalidae: Danainae) and libythea celtis (Nymphalidae: Libytheinae) and their phylogenetic implications. ISRN Genomics. 2013;**14**

[48] Kim I, Lee EM, Seol KY, et al. The mitochondrial genome of the Korean hairstreak, *Coreana raphaelis* (Lepidoptera: Lycaenidae). Insect Molecular Biology. 2006;**15**:217-225

[49] Kim MJ, Jun J, Kim I. Complete mitochondrial genome of the mulberry white caterpillar *Rondotia menciana* (Lepidoptera: Bombycidae). Mitochondrial DNA. 2014;**25**:1-3

[50] Y-P W, Zhao J-L, T-J S, et al. The complete mitochondrial genome of *Leucoptera malifoliella* Costa (Lepidoptera: Lyonetiidae). DNA and Cell Biology. 2012;**31**:1508-1522

[51] Yong-Qiang C, Ma C, Ji-Yue C, Yang DR. The complete mitochondrial genomes of two ghost moths, *Thitarodes renzhiensis* and *Thitarodes yunnanensis*: The ancestral gene arrangement in Lepidoptera. BMC Genomics. 2012;**13**:276

[52] Taanman J. The mitochondrial genome: Structure, transcription. Biochimica et Biophysica Acta (BBA): Bioenergetics. 1999;**1410**:103-123

[53] Hou LX, Ying S, Yang XW, Yu Z, Li HM, Qin X. The complete mitochondrial genome of *Papilio bianor* (Lepidoptera: Papilionidae), and its phylogenetic position within Papilionidae. Mitochondrial DNA. Epub ahead of print 2014. DOI: 10.3109/19401736.2013.873923

[54] Zhao JL, Zhang YY, Luo AR, Jiang GF, Cameron SL, Zhu C. The complete mitochondrial genome of *Spilonota lechriaspis* Meyrick (Lepidoptera: Tortricidae). Molecular Biology Reports. 2011;**38**:3757-3764

[55] Van Oppen MJH, Hislop NR, Hagerman PJ, Miller D. Gene content and organization in a segment of the mitochondrial genome of the scleractinian coral *Acropora tenuis*: Major differences in gene order within the anthozoan subclass Zoantharia. Molecular Biology and Evolution. 1999;**16**:1812-1815

[56] Zhu X-Y, Xin Z-Z, Wang Y, et al. The complete mitochondrial genome of *Clostera anachoreta* (Lepidoptera: Notodontidae) and phylogenetic implications for *Noctuoidea* species. Genomics. 2017;**109**:221-226

[57] Jeong SY, Kim MJ, Kim S-S, et al. Complete mitochondrial genome of the endangered Lycaenid butterfly *Shijimiaeoides divina* (Lepidoptera: Lycaenidae). Mitochondrial DNA Part A. 2017;**28**:242-243

[58] Zou Z, Min Q, Cheng S, et al. The complete mitochondrial genome of *Thitarodes sejilaensis* (Lepidoptera: Hepialidae), a host insect of *Ophiocordyceps sinensis* and its implication in taxonomic revision of Hepialus adopted in China. Gene. 2017;**601**:44-55

[59] Park JS, Kim MJ, Jeong SY, et al. Complete mitochondrial genomes of two gelechioids, *Mesophleps albilinella* and *Dichomeris ustalella* (Lepidoptera: Gelechiidae), with a description of gene rearrangement in Lepidoptera. Current Genetics. 2016;**62**:809-826

[60] Chen M, Tian L-L, Shi Q-H, et al. Complete mitogenome of the Lesser Purple Emperor *Apatura ilia* (Lepidoptera: Nymphalidae: Apaturinae) and comparison with other nymphalid butterflies. Dongwuxue Yanjiu. 2012;**33**:191-201

[61] Castro LR, Austin a D, Dowton M. Contrasting rates of mitochondrial molecular evolution in parasitic Diptera and Hymenoptera. Molecular Biology and Evolution. 2002;**19**:1100-1113

[62] Dyubankova N, Sochacka E, Kraszewska K, et al. Contribution of dihydrouridine in folding of the D-arm in tRNA. Organic & Biomolecular Chemistry. 2015;**13**:4960-4966

[63] Lowe TM, Eddy SR. TRNAscan-SE: A program for improved detection of transfer RNA genes in genomic sequence. Nucleic Acids Research. 1996;**25**:955-964

[64] Kim MJ, Jeong HC, Kim SR, et al. Complete mitochondrial genome of the nerippe fritillary butterfly, *Argynnis nerippe* (Lepidoptera: Nymphalidae). Mitochondrial DNA. 2011;**22**:86-88

[65] Cameron SL, Lambkin CL, Barker SC, et al. A mitochondrial genome phylogeny of Diptera: Whole genome sequence data accurately resolve relationships over broad timescales with high precision. Systematic Entomology. 2007;**32**:40-59

[66] Walker F. 'Pyralides', List of the Specimens of Lepidopterous Insects in the Collection of the British Museum London: Printed by order of the Trustees1859. p. 509-1036

[67] Chen Q, Chen Z-S, X-S G, et al. The complete mitogenome of *Parapoynx crisonalis* (Walker, 1859) (Lepidoptera: Crambidae), with phylogenetic relationships amongst three acentropine larval forms. Aquatic Insects. 2017;**38**:79-91

[68] Dowling DK, Tompkins DM, Gemmell NJ. The trojan female technique for pest control: A candidate mitochondrial mutation confers low male fertility across diverse nuclear backgrounds in *Drosophila melanogaster*. Evolutionary Applications. 2015;**8**:871-880

[69] Wolff JN, Tompkins DM, Gemmell NJ, et al. Mitonuclear interactions, mtDNA-mediated thermal plasticity, and implications for the Trojan Female Technique for pest control. Scientific Reports. 2016;**6**:30016

Chapter 7

Taxocenotic and Biocenotic Study of Lepidoptera (Rhopalocera) in Rucamanque: A Forest Remnant in the Central Valley of the Araucanía Region, Chile

Hernán Navarrete Parra and Ramón Rebolledo Ranz

Additional information is available at the end of the chapter

http://dx.doi.org/10.5772/intechopen.70924

Abstract

Considering that butterflies (Lepidoptera: Rhopalocera) are sensitive to physical and climatic changes, e.g. of temperature, humidity and solar radiation, produced by disturbances in their habitat, a survey of this group was carried out in a small remnant of native forest (Rucamanque) in the central valley of the Araucanía Region of Chile. The object was to record the composition, abundance and diversity of Rhopalocera in grassland, forest and the ecotone between them during spring and summer. The study recorded 1190 individual butterflies belonging to 25 species, 18 genera, 8 sub-families and 4 families. The highest values of species richness and abundance were obtained in the summer, of 25 species and 953 individuals; in the spring, 9 species were recorded with a total of 237 individuals. The greatest diversity and homogeneity were found in the ecotone habitat (H′=3.86; J′=0.88; λ =0.08); the values for grassland were (H′=2.73; J′=0.67; λ =0.23) and for forest (H′=2.55; J′=0.71; λ =0.23); these environments being less diverse and more homogeneous. The greatest taxocenotic similarity was found between grassland and the ecotone (54%), and the least similarity appeared between the ecotone and forest (34%). The greatest biocenotic similarity was found between the ecotone and forest (48%), and the lowest correspondence was between grassland and forest (4.18%).

Keywords: biodiversity, fragmentation, Lepidoptera, Rhopalocera, Rucamanque

1. Introduction

The Lepidoptera first appear in geological history 50 million years ago. Most have mouth-pieces adapted for sucking, although some species lack these and others have mouths adapted for mastication [1–3]. They live up to great altitudes in mountainous, forest or grassland environments. Adults feed on the juices of flowers or on dew, although some species (none resident in Chile) are attracted by putrefaction [4–6]. The Lepidoptera are the second most important order of pollinating agents, after the Hymenoptera, and may be sensitive to changes in vegetation and tree cover [5, 7]. Pollination is important for most species of flowering plants and has contributed to parallel evolution between plants and lepidopterous. It was due to the appearance of butterflies that flowers became more conspicuous and acquired increasingly attractive shapes, colours and smells [7].

The International Union for Conservation of Nature [8] states that habitat loss is the most important reason for the extinction of butterflies. In red lists of threatened species worldwide, the Lepidoptera are the next most threatened insect group after the Odonata (**Table 1**).

Butterflies are very sensitive to changes in temperature, humidity and solar radiation produced by disturbances to their habitat; inventories of their communities measuring their diversity, richness and chorological aspects therefore constitute a valid instrument for assessing their state of conservation and/or changes to the natural environment [10]. Due to their abundance, diversity, ease of handling in the field, spatial and temporal stability, and in general because their taxonomy is well documented, they are used as appropriate ecological indicators [10].

The lepidoptera have been divided [11] into *Jugatae* and *Frenatae* (Comstock, 1893); *Zeugloptera* (Chapman, 1916), *Monotrysia* (Borner, 1939) and *Ditrysia* (Borner, 1925); *Heterocera* (Boisduval, 1834) and *Rhopalocera* (Duméril, 1823). This work will consider only the latter group. The Heterocera (moths) have antennae of varying shapes, whereas the Rhopalocera (butterflies) have straight antennae ending in a small club, sometimes forming a hook, which makes them very easy to distinguish [4]. There have been few studies of the state of conservation of butterflies in Chile [12], and those that exist indicate that the Rhopalocera are poorly represented in the National System of State-Protected Wildlife Areas (SNASPE). Although central Chile is recognised as a world biodiversity hotspot, such studies are minimal in this zone and endemic species receive the least protection [13, 14].

Knowledge about conservation states is very scarce in Chile. There is a lack of data on the biology and distribution of most of the species described to date [15, 16]. Chilean species of the following genera are threatened: *Neomaenas* Wallengren, 1858; *Auca* Hayward, 1953; *Butleria*

Order	EX	CR	EN	VU	NT	DD	LC	Total
Coleoptera	16	10	16	27	3	0	0	72
Lepidoptera	27	8	39	130	45	35	19	303
Hymenoptera	0	3	0	139	7	1	1	151
Odonata	2	36	64	74	34	90	327	627
Orthoptera	2	8	8	50	0	2	0	74

Source: [9]. IUCN Red List 2007: Extinct (EX); critically endangered (CR); endangered (EN); vulnerable (VU); near threatened (NT); data deficient (DD) and least concern (LC).

Table 1. Principal groups of insects with threatened species.

Kirby, 1871; *Tatochila* Butler, 1870 and *Homoeonympha* C. & R. Felder, 1867. According to current knowledge, there are 169 butterfly species recorded for Chile, in 5 families [4] (**Table 2**).

Studies of Chilean lepidopterofauna may focus on taxocenosis or biocenosis [17–20]. In the Araucanía Region, a study done at Villarrica National Park [18] characterised the group at different altitudes; another survey was carried out on the Cerro Ñielol Natural Monument (Temuco) to obtain information for state-protected areas [17]; a third described the area of Budi Lake on the coast, which is considered a priority conservation site in the National Biodiversity Strategy [26]; and another was a study of Chilean Papilionoidea and Hesperioidea, including their flight periods [13].

Biodiversity is understood to mean the assemblage of genes, species, ecosystems and landscapes in a given space at any given moment, with their successive hierarchical interaction from genes to species, ecosystems and landscapes and vice versa [21].

Fragmentation is a dynamic process through which a habitat is reduced to small fragments or islands set in a larger area or matrix. The effect of this process is the formation of remnants of forest separated by a matrix of different kinds of vegetation or land use, the dynamics of which depend on factors occurring in the surrounding areas. It is one of the most recent causes of the impoverishment of biodiversity [21–24]. In the past, Chile's central valley from the Malleco river (36° 30′ S) to 41° S was covered by park-type native forest in the north [25] and by other formations based on *roble (Nothofagus oblicua* (Mirb.) Oerst.) and other less abundant plant species. These forests were felled or burnt to clear land for arable farming and stock-rearing [26]. A few patches of the original Roble-Laurel-Lingue formation have remained almost unaltered, by decision of some landowners. Forests of this kind, very scarce

Superfamily	Family	Subfamily	Genus	Species
Hesperioidea	Hesperiidae	Pyrginae	5	11
		Hesperiinae	7	24
Papilionoidea	Pieridae	Coliadinae	5	10
		Pierinae	8	19
	Papilionidae		1	1
	Lycaenidae	Theclinae	13	30
		Polyommatinae	6	29
	Nymphalidae	Danainae	1	1
		Satyrinae	12	34
		Heliconiinae	4	6
		Nymphalinae	2	3
		Libytheinae	1	1
Σ 2	**5**	**11**	**65**	**169**

Source: [4].

Table 2. Diurnal lepidoptera present in Chile.

but ecologically and scientifically very important, are identified as Original Remnants [36]. Today, there are only two such patches of mature native forest—typical of the original vegetation—in Cautín Province in the central valley of the Araucanía Region: Cerro Ñielol and Rucamanque. The latter is the larger of the two, covering 435.1 ha, and has suffered less human impact, with some 209 hectares covered by primary forest. The Rucamanque remnant is included in the National Biodiversity Strategy as a national priority site with *Priority II* (*Important*); its priority at the regional level of the National Strategy for the Conservation and Sustainable Use of Biodiversity is *Very High* [27]. Several studies have been carried out in the forest remnants of the Araucanía Region [28–30]; however, most deal with vegetation and very few with fauna. In view of the above, the purpose of this work is to report information on the taxonomic composition of butterflies (Rhopalocera) in Rucamanque, a forest remnant and priority conservation site.

The working hypothesis was: Considering the biotic and abiotic differences between forest, grassland and the ecotone, it is to be expected that the richness and abundance of Rhopalocera species will differ between them. The following General object was therefore proposed: To characterise the diurnal lepidopterofauna of Rucamanque, a forest remnant set in a fragmented ecosystem in the central valley of the Araucanía Region of Chile. The specific objects were: (1) to determine the taxa of diurnal lepidoptera associated with forest, grassland and ecotone environments in the Rucamanque forest remnant, determining diversity in forest, grassland and ecotone environments by community parameters and (2) to compare the taxocenotic and biocenotic similarities between the Rhopalocera of the different habitats in spring and summer using the Bray–Curtis similarity coefficient [31].

2. Materials and methods

2.1. Study area

Rucamanque (38° 39′ S, 72° 35′ W) (**Figure 1**), area 435.1 ha., is located 12 Km. NW of the city of Temuco, Cautín Province, Araucanía Region, Chile. Rucamanque forms part of the Huimpil-Ñielol range of hills. It consists of a ravine descending from NW to SE; the altitude at the foot of the ravine is approximately 200 m, and the sides climb to 550 m. The prevailing climate is wet temperate, with average annual precipitation of 1400 mm and mean temperature 12°C [31].

Field methods. A preliminary survey was done of the study area to decide on sampling areas in the three habitats: grassland (38°39′26″S, 72°36′21.72″W), ecotone (38°39′42.47″S, 72°36′18.96″W) and forest (38°39′44.56″S, 72°36′16.92″W). Specimens were sampled and recorded from the second half of September 2007 to the first half of April 2008. Sampling was carried out in the sampling area with a butterfly net on 1 day every 2 weeks (from 10:00 to 12:00, 14:00 to 16:00 and 18:00 to 19:00 hours).

The samples were recorded with capture number, habitat, activity and time, including photographs taken *in situ*. Individuals out of reach of the net were identified and counted with binoculars (7×50). Species were identified by comparison with examples from scientific collections in Temuco (Universidad de la Frontera) and Santiago (Natural History Museum, Juan

http://dx.doi.org/10.5772/intechopen.70924

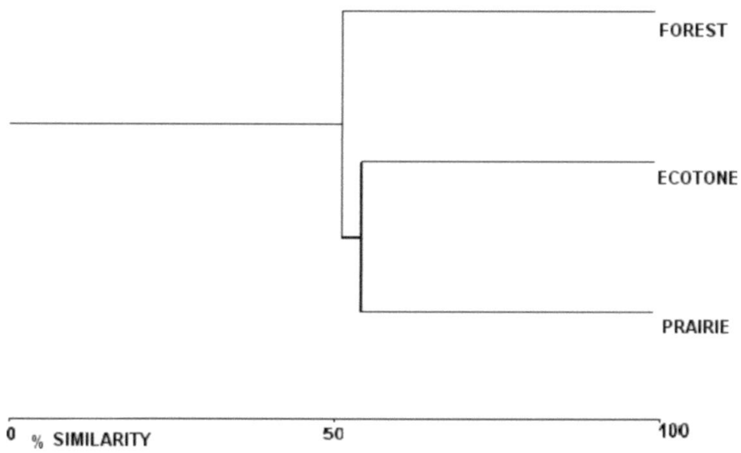

Figure 1. Taxocenotic similarity tree diagram for Rhopalocera in habitats sampled in Rucamanque, Araucanía Region.

Ignacio Molina Study and Publication Institute) and with the aid of keys [32]. The material was deposited in the Entomological Collection of the Faculty of Farming and Forestry Sciences of Universidad de la Frontera, Temuco.

Data analysis included a taxonomic list of the species present in the study area to determine the richness of species **(S)** and relative abundance **(%)** at the site to interpret the alpha diversity. Indices of diversity and abundance were obtained for each of the habitats sampled for both spring and summer seasons. The taxocenosis and biocenosis were indicated by the similarity index as proposed by Bray and Curtis [38] for all three habitats and two seasons studied (spring-summer). Tree diagrams were drawn for each result using the Biodiversity Pro programme. If the data did not fit a normal distribution, Friedman's non-parametric test was applied [34, 35].

3. Results

Taxonomic composition of species recorded. A total of 25 species was recorded among the samples (S). The families with the most representatives were the Hesperiidae, Pieridae and Nymphalidae. The family with the smallest number of samples was the Lycaenidae (**Table 3**; **Figures 7** and **8**). A new distribution was established two species: *Eiseliana bicolor* (Philippi, 1859) (Lycaenidae) and *Neomaenas coenonymphina* Butler, 1881 (Nymphalidae).

Diversity in habitats sampled: The ecotone presented the greatest richness of species (21 in total) and an abundance of 376 individuals. The grassland and forest habitats presented lower richness of species, the poorest being the forest with only 12 species and an abundance of 361 individuals. This is reflected in the diversity values (H′) for grassland, ecotone and forest—2.73, 3.86 and 2.55, respectively (**Table 4**). The habitat presenting distribution with the greatest homogeneity was the ecotone (J′ = 0.88). Grassland and forest present similar values (J′ = 0.67 and J′ = 0.71 respectively). Dominance was strongest for species of the

SUPERFAMILY	:	HESPEROIDEA	
FAMILY	:	HESPERIIDAE	
SUBFAMILY	:	PYRGINAE	Burmeister, 1878
SPECIES	:	*Pyrgus notatus valdivianus*	(Philippi)
SUBFAMILY	:	HESPERIINAE	Barnes & Benyamin, 1926
SPECIES	:	*Butleria paniscoides paniscoides*	(Blanchard, 1852)
	:	*Butleria bissexguttata*	(Philippi, 1859)
	:	*Argopteron aureipennis*	(Blanchard, 1852)
	:	*Hylephila fasciolata*	(Blanchard, 1852
	:	*Hylephila signata*	(Blanchard, 1852)
SUPERFAMILY	:	PAPILIONOIDEA	
FAMILY	:	PIERIDAE	Duponchel, 1835
SUBFAMILY	:	COLIADINAE	Swainson, 1827
SPECIES	:	*Colias vauthierii vauthierii*	Guérin, 1829
SUBFAMILY	:	PIERINAE	Swainson, 1940
SPECIES	:	*Mathania leucothea*	(Molina, 1782)
	:	*Pieris brassicae*	(Linné, 1758)
	:	*Tatochila autodice blanchardi*	(Butler, 1881)
FAMILY	:	LYCAENIDAE	Leach, 1815
SUBFAMILY	:	THECLINAE	Röber, 1892
SPECIES	:	*Strymon eurytulus*	Hübner, 1819
	:	*Eiseliana bicolor*	(Philippi, 1859)
FAMILY	:	NYMPHALIDAE	Swainson, 1837
SUBFAMILY	:	SATYRINAE	Boisduval, 1833
SPECIES	:	*Cosmosatyrus chilensis chilensis*	(Guérin, 1832)
	:	*Homoeonympha boisduvali pusilla*	(C. & R. Felder 1852)
	:	*Homoeonympha humilis*	(C. & R. Felder, 1857)
	:	*Neomaenas coenonymphina*	
	:	*Neomaenas janirioides*	Blanchard, 1852.
	:	*Neomaenas monachus*	(Blanchard, 1852)
	:	*Neomaenas poliozona*	(C. & R. Felder, 1867)
	:	*Auca pales*	(Philippi, 1859)

http://dx.doi.org/10.5772/intechopen.70924

	:	*Elina montroli*	(Feisthamel, 1839)
	:	*Nelia nemyroides*	(Blanchard, 1852)
SUBFAMILY	:	HELICONIINAE	Swainson, 1887
SPECIES	:	*Yramea cytheris*	(Drury, 1773)
SUBFAMILY	:	NYMPHALINAE	Swainson, 1827
SPECIES	:	*Vanessa carye*	(Hübner, 1806)
	:	*Vanessa terpsichore*	Philippi, 1859.

Table 3. Systematic list of species of Lepidoptera—Rhopalocera recorded in Rucamanque.

Community parameters	Sampling stations		
	Grassland (E1)	Ecotone (E2)	Forest (E3)
Abundance (N)	453	376	361
Species richness (S)	17	21	12
Simpson's dominance index (λ)	0.23	0.08	0.23
Shannon-Wiener diversity index (H′)	2.73	3.86	2.55
Theoretical maximum index (H′ Max.)	4.09	4.39	3.58
Homogeneity (J′)	0.67	0.88	0.71

Table 4. Community parameters analysed in Rucamanque.

grassland (*Colias vauthierii vauthierii* and *Hylephila fasciolata*) and forest (*Neomaenas monachus* and *Argopteron aureipennis*); the ecotone, which presented greater equity, as a result presented the lowest dominance $\lambda = 0.08$ (**Table 4**).

The greatest richness of species was recorded in summer, when 21 of the 25 species recorded in this study were reported, with an abundance of 953 individuals. Spring presented less diversity, with 17 species recorded and an abundance of 237 individuals.

The greatest homogeneity was found in the ecotone, in both spring and summer. This is reflected in the diversity (H′) and homogeneity (J′) obtained in each environment and season (**Table 5**).

Taxocenotic and biocenotic similarity. Regarding to the presence or absence of species in the habitats sampled (**Table 6**), *Pyrgus notatus valdivianus*, *Eiseliana bicolor* and *Cosmosatyrus ch. chilensis* are found in grassland, *Pieris brassicae* is found in the ecotone, whereas species like *Mathania leucothea*, *Homoeonympha boisduvali pusilla*, *Neomaenas monachus* and *Vanessa carye* are found in all three habitats. The greatest number of shared species were shared between grassland and ecotone. The taxocenotic similarity between habitats is shown in **Figure 1**.

Community parameters	Spring 2007			Summer 2008		
	P	E	B	P	E	B
Total number of individuals (N)	136	59	42	317	317	319
Species richness (S)	8	6	3	15	21	12
Shannon-Wiener diversity index (H′)	2.05	2.1	1.27	2.42	3.71	2.28
Theoretical H′ index (H′ Max.)	3	2.58	1.8	3.91	4.39	3.58
Homogeneity (J′)	0.68	0.81	0.8	0.62	0.84	0.64

Table 5. Community parameters obtained in grassland (P), ecotone (E) and forest (B) in spring 2007 and summer 2008.

Species	Grassland	Ecotone	Forest
Pyrgus notatus valdivianus	P	A	A
Butleria paniscoides paniscoides	A	P	P
Butleria bissexguttata	A	P	P
Argopteron aureipennis	A	P	P
Hylephila fasciolata	P	A	P
Hylephila signata	P	P	A
Colias vauthierii vauthierii	P	P	A
Mathania leucothea	P	P	P
Pieris brassicae	A	P	A
Tatochila autodice blanchardi	P	P	A
Strymon eurytulus	P	P	A
Eiseliana bicolor	P	A	A
Cosmosatyrus chilensis chilensis	P	A	A
Homoeonympha boisduvali pusilla	P	P	P
Homoeonympha humilis	A	P	P
Neomaenas coenonymphina ssp.	A	P	P
Neomaenas janirioides	P	P	A
Neomaenas monachus	P	P	P
Neomaenas poliozona	A	P	P
Auca pales	P	P	A
Elina montroli	P	P	A
Nelia nemyroides	A	P	P
Yramea cytheris	P	P	A
Vanessa carye	P	P	P
Vanessa Terpsichore	P	P	A

Table 6. Presence (P) and absence (A) of Rhopalocera species recorded by habitat in spring 2007 and summer 2008 in Rucamanque, Araucanía Region.

Taxocenotic similarity matrix	Grassland	Ecotone	Forest
Grassland	*	54.16	34.48
Ecotone	*	*	51.16
Forest	*	*	*

Table 7. Taxocenotic similarity matrix for Rhopalocera in habitats sampled in Rucamanque, Araucanía Region.

Species	Grassland		Ecotone		Forest		Total	
	N	%	N	%	N	%	N	%
Pyrgus notatus valdivianus	3	0.66	0	0.00	0	0.00	3	0.25
Butleria paniscoides paniscoides	0	0.00	5	1.33	8	2.22	13	1.09
Butleria bissexguttata	0	0.00	35	9.31	40	11.08	75	6.30
Argopteron aureipennis	0	0.00	16	4.26	120	33.24	136	11.43
Hylephila fasciolata	126	27.81	0	0.00	2	0.55	128	10.76
Hylephila signata	22	4.86	4	1.06	0	0.00	26	2.18
Colias vauthierii vauthierii	164	36.20	39	10.37	0	0.00	203	17.06
Mathania leucothea	11	2.43	15	3.99	3	0.83	29	2.44
Pieris brassicae	0	0.00	6	1.60	0	0.00	6	0.50
Tatochila autodice blanchardi	9	1.99	2	0.53	0	0.00	11	0.92
Strymon eurytulus	2	0.44	2	0.53	0	0.00	4	0.34
Eiseliana bicolor	1	0.22	0	0.00	0	0.00	1	0.08
Cosmosatyrus chilensis chilensis	37	8.17	0	0.00	0	0.00	37	3.11
Homoeonympha boisduvali pusilla	30	6.62	37	9.84	1	0.28	68	5.71
Homoeonympha humilis	0	0.00	25	6.65	35	9.70	60	5.04
Neomaenas coenonymphina ssp.	0	0.00	32	8.51	16	4.43	48	4.03
Neomaenas janirioides	4	0.88	36	9.57	0	0.00	40	3.36
Neomaenas monachus	1	0.22	58	15.43	109	30.19	168	14.12
Neomaenas poliozona	0	0.00	11	2.93	2	0.55	13	1.09
Auca pales	17	3.75	6	1.60	0	0.00	23	1.93
Elina montroli	2	0.44	20	5.32	0	0.00	22	1.85
Nelia nemyroides	0	0.00	13	3.46	15	4.16	28	2.35
Yramea cytheris	4	0.88	9	2.39	0	0.00	13	1.09
Vanessa carye	18	3.97	4	1.06	10	2.77	32	2.69
Vanessa terpsichore	2	0.44	1	0.27	0	0.00	3	0.25
Σ habitats and study area	**453**	100.00	**376**	100.00	**361**	100.00	**1190**	100.00

Table 8. Absolute (N) and relative (%) abundance by species in the study area.

The taxocenotic similarity between the habitats sampled shows greater correspondence between ecotone and grassland, forming a nucleus; this nucleus corresponds to a lesser degree with the forest habitat (**Table 7**; **Figure 1**).

Grassland and ecotone present the highest abundances, with 453 and 376 individuals recorded, respectively (**Table 8**). In the forest, 361 individuals were recorded.

The most abundant species were *Colias v. vauthierii*, *Neomaenas monachus*, *Hylephila fasciolata* and *Argopteron aureipennis* with 203, 163,128 and 122 individuals, respectively. Low abundance, not exceeding six individuals over the sampling period, was found in *Pyrgus notatus valdivianus*, *Mathania leucothea*, *Strymon eurytulus*, *Eiseliana bicolor* and *Vanessa terpsichore*.

Figure 5 shows the biocenotic similarity of the habitats studied. In general, the similarity between habitats is low; the highest correspondence exists between the ecotone and the forest, forming a nucleus with 48% similarity; however, these two environments together are more distant from grassland, which presents less similarity (**Table 9**; **Figure 2**).

The taxocenotic and biocenotic similarities differed between spring and summer, due to the larger number of species and greater abundances recorded in summer.

Spring presented a higher taxocenotic similarity nucleus (66%) between ecotone and forest. There was a high level of correspondence between grassland and ecotone (57%). The lowest correspondence was found between grassland and forest (**Table 10**; **Figure 3**).

Biocenotic similarity matrix	Grassland	Ecotone	Forest
Grassland	*	26.54	4.18
Ecotone	*	48.30	
Forest	*	*	*
Total study area	453	376	361

Table 9. Biocenotic similarity matrix for Rhopalocera in the habitats sampled in Rucamanque, Araucanía Region.

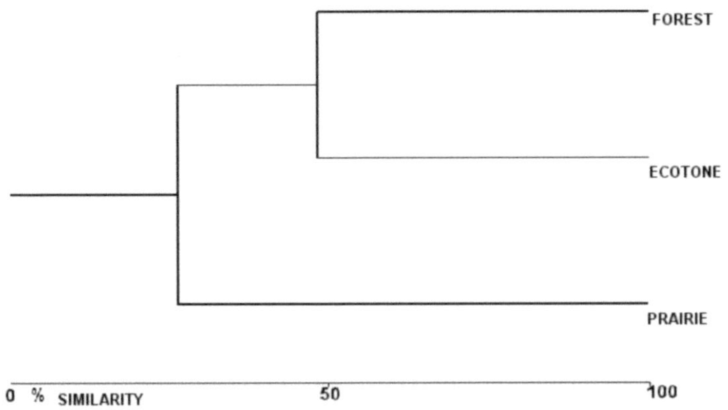

Figure 2. Biocenotic similarity tree diagram for Rhopalocera in the habitats sampled in Rucamanque, Araucanía Region.

Taxocenotic similarity matrix, spring 2007

	Spring-Forest	Spring-Ecotone	Spring-Grassland
Spring-Forest	*	66.67	18.18
Spring-Ecotone	*	*	57.14
Spring-Grassland	*	*	*

Table 10. Taxocenotic similarity matrix for Rhopalocera in Rucamanque, Araucanía Region, spring 2007.

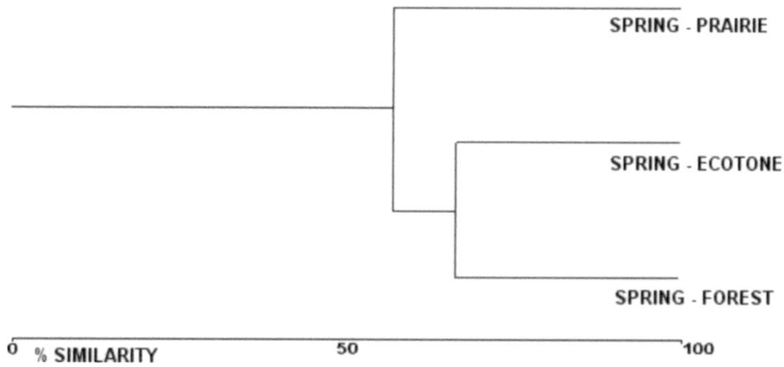

Figure 3. Taxocenotic similarity tree diagram for Rhopalocera in Rucamanque, Araucanía Region, spring 2007.

In summer, the taxocenotic similarity forms a single nucleus. There was no change in the percentages between ecotone and forest as compared to spring; however, there was an increase in correspondence between grassland-ecotone and grassland-forest. From this, we infer that the difference may result from records of "tourist" species observed primarily in the grassland environment, which seek cooler, more stable conditions during the heat of the day; as a result, they are sampled in environments to which they do not strictly belong or do not habitually visit (**Table 11**; **Figure 4**).

Biocenotic similarity in spring (**Figure 5**), strong correspondence is observed between the ecotone and the forest, forming a nucleus with 61% similarity, followed by grassland-ecotone with 23%.

In summer, the biocenotic similarity between the habitats is low, not exceeding 50%. The greatest correspondence is found in the nucleus which combines ecotone and forest, whereas the lowest similarity is between grassland and forest (**Tables 12** and **13**; **Figure 5**).

Taxocenotic similarity matrix, summer 2008

	Summer-Forest	Summer-Ecotone	Summer-Grassland
Summer-Forest	*	66.67	29.63
Summer-Ecotone	*	*	66.67
Summer-Grassland	*	*	*

Table 11. Taxocenotic similarity matrix for Rhopalocera in Rucamanque, Araucanía Region, summer 2008.

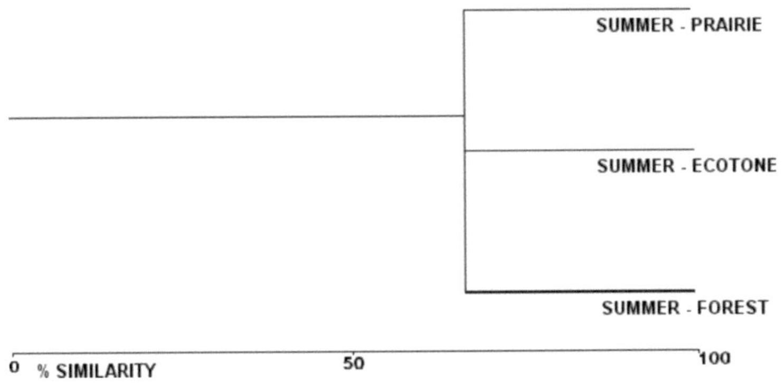

Figure 4. Taxocenotic similarity tree diagram for Rhopalocera in Rucamanque, Araucanía Region, summer 2008.

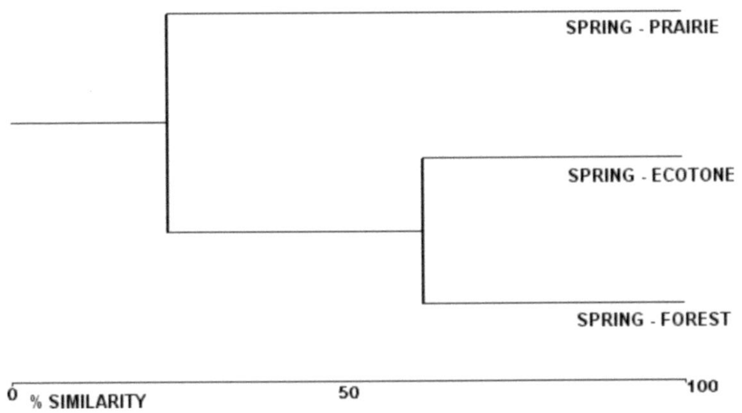

Figure 5. Biocenotic similarity tree diagram for Rhopalocera in Rucamanque, Araucanía Region, spring 2007.

Biocenotic similarity matrix, spring 2007

	Spring-Forest	Spring-Ecotone	Spring-Grassland
Spring-Forest	*	61.39	4.49
Spring-Ecotone	*	*	23.59
Spring-Grassland	*	*	*
Total study area	42	59	136

Table 12. Biocenotic similarity matrix for Rhopalocera in Rucamanque, Araucanía Region, spring 2007.

Statistical analysis with Friedman's test (Spieguel, 2002) ($p \leq 0.05$) shows that the greatest significant difference is recorded between ecotone and forest, with a smaller similar difference between these two environments and grassland. This means finally that there is a difference between the environments studied (**Figures 5** and **6**).

Biocenotic similarity matrix, summer 2008

	Summer-Forest	Summer-Ecotone	Summer-Grassland
Summer-Forest	*	45.6	3.77
Summer-Ecotone	*	*	23.03
Summer-Grassland	*	*	*
Total study area	319	317	317

Table 13. Biocenotic similarity matrix for Rhopalocera in Rucamanque, Araucanía Region, summer 2008.

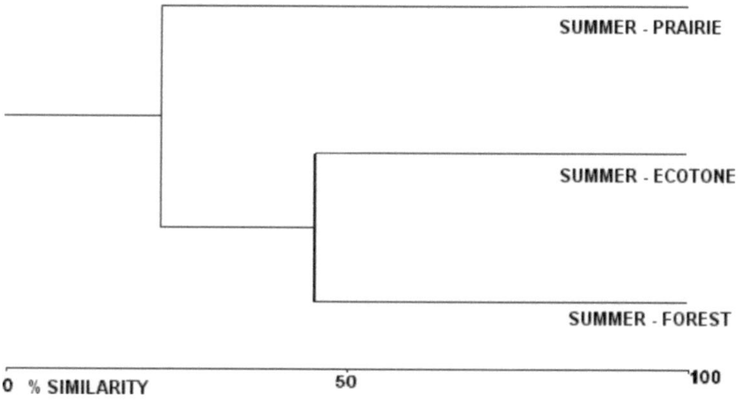

Figure 6. Biocenotic similarity tree diagram for Rhopalocera in Rucamanque, Araucanía Region, summer 2008.

The community parameters calculated and the Bray & Curtis similarity coefficient [33] denote these differences.

4. Discussion

From the data obtained in the study, we can indicate that the greatest richness of species is found in the ecotone, with 21 species; this habitat also presents the least dominance and highest equity in its population structure. It may be inferred that the least diverse habitat is forest, with only 12 species and the lowest homogeneity due to the presence of two dominant species: *Argopteron aureipennis* and *Neomaenas monachus*. Grassland presented greater diversity with 17 species; however, there were also two dominant species: *Colias vauthierii vauthierii* and *Hylephila fasciolata,* which together made up 52% of the population in this environment.

The absolute abundance, considering the whole sampling period, was highest in grassland with 453 individuals, followed by ecotone with 376 and finally forest with 361. These results agree with the results of Concha–Bloomfield et al. [21] in similar environments.

The differences in diversity between seasons result from abiotic factors [21], such as sunlight and stable temperatures, and biotic factors, such as the presence of flowers. These factors are neither stable nor predominant in spring or at least in early spring. According to Ollerton [8], the lepidoptera are in general the order of insects most closely associated with flowering plants and most dependent on flowers as a food source. This explains the greater richness of species recorded in the summer, when 21 of the 25 species recorded in the study area were observed.

There were marked differences between the abundances in different seasons. In summer, a total of 953 individuals was recorded, whereas in spring the total was 237—respectively 80.08 and 19.92% of the total individuals reported in the study. This difference may result from the better environmental characteristics, which favour hatching in the majority of species. Adults of 9 of the 25 species recorded in the study area were observed in flight in spring, with high dominance of a few species, such as *Hylephila fasciolata, Hylephila signata, Homoeonympha boisduvali pusilla* and *Homoeonympha humilis.*

In summer, the majority of species were observed in flight, with the greatest abundances in January and February. This observation appears to be characteristic of different altitudes and is related with good weather conditions, since in the same region, at an average altitude of 1345 m.a.s.l. Parada et al. [20] observed the largest number of specimens at this time of year. Abiotic factors like humidity and temperature, dependent on sunlight, and biotic factors like flowering, influence the hatching and establishment of the species found in Rucamanque during this season, corroborating [21].

Some species were only collected in one specific month during the sampling period: *Pyrgus notatus valdivianus* (January), *Butleria paniscoides paniscoides* (February), *Pieris brassicae* (March) and *Cosmosatyrus chilensis chilensis* (February). Likewise, there are species, such as *Pyrgus notatus valdivianus, Eiseliana bicolor* and *Cosmosatyrus ch. chilensis,* which are found exclusively in grassland, and *Pieris brassicae* recorded only in the ecotone. This may be because some species require a specific host plant for feeding or laying their eggs.

The taxocenotic similarity tree diagram indicates great similarity between grassland and ecotone, forming a nucleus of 54%; these two environments form a second nucleus at 51%. The lowest similarity, between forest and grassland, was 34%. These values are explained by the fact that grassland and ecotone share a large number of species, given the favourable environmental conditions present in these sectors such as greater amount of solar radiation and greater number of flowers. On the other hand, forest and grassland are exclusive home to some species, depending on their natural characteristics; these species tend to be more specific and less generalist.

The biocenotic similarity tree diagram indicates the existence of a similarity nucleus of 48% between ecotone and forest. The ecotone in turn has a similarity of 26% to grassland. Again, the environments with the lowest similarity are grassland and forest. The only slight differences between different habitats may be influenced by tourist species, which seek out more favourable environmental conditions when solar radiation is strong, and are recorded in the survey although they do not reside exclusively in this environment. This is a determining factor, since although the influence of environmental factors triggered by sunshine is higher in open areas [11], this migration may increase when the area has suffered greater alteration, as in the case of Rucamanque.

1
30 mm
2
22 mm
3
23 mm
4a
30 mm
4b
5
23 mm
6
45 mm
7
31 mm
8
53 mm
9
51 mm
10a
43 mm
10b
11
12a
21 mm
12b
13
a
50 mm
b
14
a
29 mm
b
15
a
30 mm
b
16
a
37 mm
b
17
a
43 mm
b

Figure 7. Dorsal and ventral views of the species recorded (adults): (1) *Pyrgus n. valdivianus* dorsal; (2) *Butleria p. paniscoides* dorsal; (3) *Butleria bissexguttata* dorsal and ventral; (4a) *Argopteron aureipennis* dorsal; (4b) *Argopteron aureipennis* ventral; (5) *Hylephila fasciolata* dorsal and ventral; (6) *Colias v. vauthierii* ♂ and ♀ dorsal; (7) *Hylephila signata* dorsal and ventral; (8) *Mathania leucothea* dorsal; (9) *Pieris brassicae* dorsal; (10a) *Tatochila autodice blanchardi* dorsal; (10b) *Tatochila autodice blanchardi* ventral; (11) *Strymon eurytulus* lateral; (12a) *Eiseliana bicolor* dorsal; (12b) *Eiseliana bicolor* ventral; (13) *Cosmosatyrus ch. chilensis* dorsal (a) and ventral (b); (14) *Homoeonympha b. pusilla* dorsal (a) and ventral (b); (15) *Neomaenas coenonymphina* dorsal (a) and ventral (b); (17) *Neomaenas janirioides* dorsal and ventral; (18) *Neomaenas monachus* dorsal (a) and ventral (b); (19) *Elina montroli* dorsal (a) and ventral (b); (20a) *Nelia nemyroides* dorsal; (20b) *Nelia nemyroides* ventral; (21) *Auca pales* dorsal (a) and ventral (b); (22) *Neomaenas poliozona* dorsal (a) and ventral (b); (23) *Yramea cytheris* dorsal (a) and ventral (b); (24) *Vanessa carye* dorsal (a) and ventral (b); (25a) *Vanessa terpsichore* dorsal; (25b) *Vanessa terpsichore* ventral.

Turning to the taxocenotic similarity between seasons, in spring, there was greater similarity between ecotone and forest (66.67%), and these two environments together diverge from grassland in summer, which has a lower similarity. This occurs because the majority of the species recorded in grassland in spring are not shared, a large number being found exclusively in this habitat.

The biocenotic similarity does not present great differences. The greatest dissimilarity is found between forest and grassland; this is explained by the fact that these two environments share fewer species; however, spaces which have suffered intervention and disturbance have higher levels of light [14] causing a restructuring of the butterfly community around forest edges. This allows pioneer species to increase in the plant succession, increasing food resources.

http://dx.doi.org/10.5772/intechopen.70924

Figure 8. Photographs of Rhopalocera taken "*in situ*" at the Rucamanque site, spring 2007–summer 2008, on possible host plants: (1) *Pyrgus n. valdivianus* (Philippi) (Ea =; (2) *Butleria p. paniscoides* (Blanchard, 1852); (3) *Butleria bissexguttata* (Philippi, 1859); (4) *Argopteron aureipennis* (Blanchard, 1852); (5) *Hylephila fasciolata* (Blanchard, 1852); (6) *Mathania leucothea* (Molina, 1782); (7) *Pieris brassicae* (Linné, 1758); (8) *Tatochila a. blanchardi* (Butler, 1881); (9) *Strymon eurytulus* Hubner, 1819.; (10) *Eiseliana bicolor* (Philippi, 1859); (11) *Homoeonympha boisduvali pusilla* (C. & Felder); (12) *Cosmosatyrus chilensis chilensis* (Guérin, 1832): (13) *Homoeonympha humilis* (C. & R. Felder, 1867; (14) *Neomaenas monachus* (Blanchard, 1852); (15) *Neomaenas poliozona* (C. & R. Felder, 1867); (16) *Nelia nemyroides* (Blanchard, 1852); (17) *Elina montroli* (Feisthamel, 1839); (18) Yramea *cytheris* (Drury, 1773); (19) *Auca pales* (Philippi, 1859); (20) *Neomaenas coenonymphina ssp.*; (21) *Vanessa terpsichore* Philippi, 1859; (22) *Vanessa carye* (Hübner, 1806); (23) *Neomaenas janirioides* Blanchard, 1852; (24a) *Hylephila signata* (♂) (Blanchard, 1852). (24b) *Hylephila signata* (♀m) (Blanchard, 1852); (25a) *Colias vauthierii vauthierii* (♂) Guérin 1829.; (25b) *Colias vauthierii vauthierii* (♀) Guérin 1829.

Of particular interest in this study are the species *Eiseliana bicolor* (Philippi, 1859) and *Neomaenas coenonymphina* Butler, 1881, for which the distribution range must be extended. According to Ref. [4], the distribution of *E. bicolor* is from Atacama to the Bío-Bío. *N. coenonymphina* is an uncommon species supposedly found only in Chile's fifth region; it is associated with the graminid *Chusquea sp.* Kunth; this record is the more important as the species is considered to be threatened [17].

The forest remnant Rucamanque is the most pristine part of the Huimpil-Ñielol range of hills. It is therefore strange that the important indicator species *Eroessa chilensis* (Guérin, 1830) was not recorded and probably results from sampling errors.

In earlier studies in this area, Cerro Ñielol, close to the centre of Temuco, presented the same number of species [26]; it should however be noted that this study is 25 years old. Other studies reporting the diversity of Rhopalocera at different altitudes in the Araucanía Region [20, 26] indicate that the greatest diversity is found in the central valley and in forest remnants like Rucamanque, identified with the subtype name Original Remnants [36]. Paradoxically, this sector is suffering severe anthropic pressure, since although it is a priority site for conservation, it enjoys no legal protection as does the Cerro Ñielol Natural Monument.

The greatest diversity and levels of endemicism for Rhopalocera are found in central Chile. This zone is recognised as a biodiversity hotspot, for both the endemicism and richness of its vegetation [21]. This observation is also applicable to invertebrates, including insects, since plant diversity patterns are frequently good predictors of insect diversity patterns. The presence of certain species of Rhopalocera indicates that its host plants are present nearby [21].

The diversity of Rhopalocera present in the Rucamanque forest remnant is the more important because of the fragmentation process that is continuing in south central Chile (**Figures 7** and **8**).

The ecological characteristics of the Rucamanque remnant are still predominantly stable, despite fragmentation and the exotic plantations in the vicinity. They have allowed a great diversity of lepidoptera to flourish, including 50% of the species recorded for the Araucanía Region and 14% of the species recorded in Chile. According to Donoso [36, 37], forests of this kind are very scarce, but ecologically and scientifically very important. This explains the importance of this priority site for the conservation of biodiversity and the need to survey the micro and macrofauna in this sector.

5. Conclusions

- Study of the Rucamanque forest remnant produced a total of 25 species of Rhopalocera with a total of 1190 individuals recorded, representing 4 families and 18 genera of the order Lepidoptera. These 25 species represent 50% of the local lepidoptera and 14.79% of those recorded in Chile.
- The habitat where the greatest diversity of species was found was the ecotone, where 21 species were recorded representing more than 80% of all the species reported at the study site.

- The habitat presenting the least diversity was the forest with 12 species, 48% of all the species recorded in the study area.
- The seasonal diversity and abundance were stronger in summer, with higher values for species diversity and abundance in all three habitats sampled.
- The greatest numbers of species shared between habitats were 12 between ecotone and grassland and 11 between ecotone and forest. Grassland and forest presented the smallest number of shared species, only 4. The greatest taxocenotic similarity was found between ecotone and grassland, while forest presented the lowest similarity, diverging from these two habitats.
- The greatest abundance occurred in the grassland, with 453 individuals; the biocenotic similarity of grassland with ecotone and forest was low and the similarity between these two was higher. Although a larger number of species share the grassland and ecotone habitats, two species dominate the grassland assemblage, accounting for 64% of the population in this habitat, meaning that the other species are present in smaller numbers.
- New distributions were recorded for the species *Eiseliana bicolor* (Philippi, 1859) and *Neomaenas coenonymphina* Butler, 1881.
- Statistically significant differences were found in the three environments studied in Rucamanque, the greatest differences occurring between forest and grassland. The community parameters calculated and the Bray & Curtis similarity coefficient denote these differences.
- Recording the species diversity of the zoological group studied in forest remnants is of great importance, both for their ecological value and to take informed decisions in management and planning for conservation, study and tourism.

Author details

Hernán Navarrete Parra[1] and Ramón Rebolledo Ranz[2]*

*Address all correspondence to: ramón.rebolledo@ufrontera.cl

1 Facultad de Recursos Naturales, Universidad Católica de Temuco, Escuela de Ciencias Ambientales, Temuco, Chile

2 Faculty of Agricultural and Forest Sciences, University of La Frontera, Temuco, Chile

References

[1] Coronado R, Márquez A. Introducción a la Entomología morfología y taxonomia de los insectos. S.A. México.: Limusa-Wiley; 1972. 282 pp

[2] Peña L. Introducción al Estudio de los Insectos de Chile. Santiago, Chile: Editorial Universitaria. 1996. 253 pp

[3] Benyamini D, Ugarte A, Shapiro A, Hermann O, Pycrz T, Bálini Z. An updated list of butterflies of Chile (Lepidoptera, Papilionoidea and Hesperoidea) including distribution, flight period and conservation status. Boletín del Museo de Historia Natural, Chile. 2014; **63**:9-31

[4] Peña L, Ugarte A. Las Mariposas de Chile. Editorial Universitariax. Editorial Universitaria, Chile. 1997. 359 pp

[5] McGAvin G. Entomología Esencial. Ariel Ciencia. Barcelona, España. 2002. 350 p

[6] Ugarte A. 2015 Pequeña guía de Campo. Mariposas de Chile. Santiago, Chile: Ograma Impresores; 250 p

[7] Lamas G. Estado actual del conocimiento de la Sistemática de los Lepidópteros, con especial referencia a la región neotropical. In: Proyecto CYTED para el Inventario y Estimación de la Diversidad Entomológica en Ibero América. Lima, Perú: Departamento de Entomología Museo de Historia Natural Universidad Nacional Mayor de San Marcos; 2000

[8] Ollerton J. Evolución y filogenia de Artropoda. Seccion V: Ecología Evolutiva. La evolución de las relaciones polinizador-planta. Boletin Sociedad Entomológica Aragonesa. 1999;**26**:741-758

[9] Prieto M, Constantino L. Abundancia, distribución y diversidad de mariposas (Lep. Rhopalocera) enel Río Tatabro, Buenaventura (Valle-Colombia). Bol. Museo Entomológico Universidad del Valle. 1996;**4**(2):11-18

[10] Mielke O, Casagrande M. Catalogo bibliográfico dos taxa superiores da ordem lepidoptera. Revista Brasileira de Zoología. 2006;**23**(1):1-41

[11] Tobar D, Rangel J, Andrade M. Diversidad de mariposas (Lepidoptera: Rhopalocera) en la parte alta de la cuenca del Río El Roble (Quindío- Colombia). Caldasia. 2002;**24**(2):393-409

[12] Comisión Nacional del MEdio ambiente CONAMA (eds). Estrategia Regional de Conservación y uso Sustentable de la Biodiversidad. Desarrollado en el marco de la elaboración de la Estrategia Nacional de Conservación y uso Sustentable de la Biodiversidad Región de la Araucanía. Chile; 2003. 170 p

[13] Kremen C, Colwell R, Erwin T, Murphy D, Noss R, Sanjayan M. Terrestrial arthropod assemblages: Their use in conservation planning. Conservation Biology. 1993;**7**(4):796-808

[14] Tobar D. Efecto de hábitat Sobre la Comunidad de Mariposas Diurnas en Un Paisaje Fragmentado del norte de Costa Rica. Centro Agronómico Tropical de Investigación y Enseñanza CATIE. Turrialba, Costa Rica. 2004. 87 pp

[15] Parra LR. Villagrán Orden Lepidoptera (Mariposas) en Rovira, J; Ugalde, J. y M. Sturin Biodiversidad de Chile. Patrimonio y desafíos. COMISIÓN NACIONAL DEL MEDIO AMBIENTE. Tercera Edición Actualizada, Editorial Ocho libros. Chile: Santiago; 2008. pp. 156-166

[16] Simonetti J, Arroyo M, Spotorno A, Lozada E, editors. Diversidad biológica de Chile. Santiago, Chile: Comité Nacional de Diversidad Biológica. Comisión Nacional de Investigación Científica y Tecnológica; 1995. 364 pp

[17] Herrera J. La vida silvestre: ¿Se extinguen las mariposas en Chile? Academia Superior de Ciencias Pedagógicas de Santiago. Rev. Academia. Chile. 1982;**3**:203-232

[18] Etcheverry M, Herrera J. Curso teórico práctico de Entomología. Editorial Universitaria. Santiago, Chile. 1972. 385 pp

[19] Horn K, Medina M. Contribución al conocimiento de la entomofauna del Parque Nacional Cerro Ñielol. Tesis de grado. Chile: Biblioteca Universidad Católica de Temuco; 1981

[20] Parada EL, Peredo G, Cabrera S, Alvares O. Mariposas diurnas del Parque Nacional Villarrica un estudio taxocenótico y biocenotico en los distritos Rucapillan y Quetrupillan. Bol.Soc.Biol. Concepción.Chile. (1993)Tomo;**64**:163-170

[21] Concha-Bloomfield I, Marquet PA, Barría I, Parra L. Análisis de representatividad de las mariposas diurnas de Chile y prioridades geográficas para su conservación. Departamento de Zoología, Facultad de Ciencias Naturales y Oceanográficas, Universidad de Concepción; Centro de Estudios Avanzados en Ecología y Biodiversidad y Departamento de Ecología, Facultad de Ciencias Biológicas, Pontificia Universidad Católica de Chile; 2004. pp. 0-5

[22] Concha-Bloomfield I, Parra, L. Análisis cualitativo y cuantitativo de la diversidad de mariposas de la Estación Biológica Senda Darwin, Chiloé, X Región, Chile, Chile, Gayana. 2006;**70**(2):186–119

[23] Batary P, Baldi A. Evidence of an edge effect on avian nest success. Conservation Biology. 2004;**18**:389-400

[24] Figueroa E, Simonetti J. Globalización y Biodiversidad: Oportunidades y desafíos para la sociedad chilena. Santiago, Chile: Editorial Universitaria; 2003. 327 pp

[25] Hernández B, Maes J, Harvey C, Vilches S, Medina A, Sanchez D. Abundancia y diversidad de escarabajos coprófagos y mariposas diurnas en un paisaje ganadero en el departamento de Rivas, Nicaragua. Agroforestería en las The Americas. 2003;**10**:39-40

[26] Hernandez O, Mendoza D. Taxocenosis lepidópterológica presente en la riberas del Lago Budi IX Región Chile [Tesis de grado]. Chile: Biblioteca Universidad Católica de Temuco; 1986

[27] Di Castri F, HAYEK E. Bioclimatologia de Chile. Ediciones Universidad Católica de Chile, Santiago, Chile. 1976. 127 pp

[28] Hauenstein E, Ramírez C, Latsague M, Contreras D. Origen fitogeográfico y espectro biológico como medida del grado de intervención antrópica en comunidades vegetales. Medio Ambiente. 1988;**9**(1):140-142

[29] Pisano E. Fitogeografía. La vegetación de las distintas zonas geográficas chilenas. Revista Geográfica de Chile Terra Australis. 1954;**11**:95-107

[30] Ramírez C, San Martín C. Estudio fitosociológico de la vegetación pratense del valle del río Chol-Chol (Cautin, Chile). Agro Sur. 1991;**22**(1):41-56

[31] Salas C. Caracterización básica del relicto de biodiversidad Rucamanque. Bosque Nativo. 2001;**29**:3-9

[32] Salas C. Ajuste y validación de ecuaciones de volumen para un relicto del bosque de Roble-Laurel-Lingue. Bosque (Valdivia). 2002;**23**(2):81-92

[33] Saunders D, Hobbs R, Margukes C. Biological consequences of ecosystem fragmentation: A review. Conservation Biology. 1991;**5**:18-31

[34] Spieguel M, Stephens R. Estadística. México: Editorial Mcgraw-Hill; 2002. 564 pp

[35] International Union For Conservation Of Nature And Natural Resources UICN www. Red List – Summary Statistics for Globally Threatened Species.Mht. 2007

[36] Donoso C. Tipos Forestales de los Bosques Nativos de Chile. Investigación y Desarrollo Forestal (CONAF/PNUD/FAO) Documento de Trabajo N38. Santiago, Chile: Publicación FAO Chile; 1981. 82 pp

[37] Donoso C. Modificaciones del paisaje forestal chileno a lo largo de la historia. Encuentro Científico Medio Ambiente Chileno. Versiones Abreviadas 1983;**1**:109–113

[38] Bray J, Curtis J. An ordination of upland forest communities of Southerm Wiscosin. Ecological Monographa. 1957;**27**:325-349